These safety symbols are used in laboratory and field investigations in this book t
ing of each symbol and refer to this page often. *Remember to wash your hands t*

M000092167

PROTECTIVE EQUIPMENT   Do not begin any lab without the proper protection equipment.

GOGGLES	Proper eye protection must be worn when performing or observing science activities that involve items or conditions as listed below.	
APRON	Wear an approved apron when using substances that could stain, wet, or destroy cloth.	
SOAP	Wash hands with soap and water before removing goggles and after all lab activities.	
GLOVES	Wear gloves when working with biological materials, chemicals, animals, or materials that can stain or irritate hands.	

LABORATORY HAZARDS

Symbols	Potential Hazards	Precaution	Response
DISPOSAL	contamination of classroom or environment due to improper disposal of materials such as chemicals and live specimens	• DO NOT dispose of hazardous materials in the sink or trash can. • Dispose of wastes as directed by your teacher.	• If hazardous materials are disposed of improperly, notify your teacher immediately.
EXTREME TEMPERATURE	skin burns due to extremely hot or cold materials such as hot glass, liquids, or metals; liquid nitrogen; dry ice	• Use proper protective equipment, such as hot mitts and/or tongs, when handling objects with extreme temperatures.	• If injury occurs, notify your teacher immediately.
SHARP OBJECTS	punctures or cuts from sharp objects such as razor blades, pins, scalpels, and broken glass	• Handle glassware carefully to avoid breakage. • Walk with sharp objects pointed downward, away from you and others.	• If broken glass or injury occurs, notify your teacher immediately.
ELECTRICAL	electric shock or skin burn due to improper grounding, short circuits, liquid spills, or exposed wires	• Check condition of wires and apparatus for fraying or uninsulated wires, and broken or cracked equipment. • Use only GFCI-protected outlets	• DO NOT attempt to fix electrical problems. Notify your teacher immediately.
CHEMICAL	skin irritation or burns, breathing difficulty, and/or poisoning due to touching, swallowing, or inhalation of chemicals such as acids, bases, bleach, metal compounds, iodine, poinsettias, pollen, ammonia, acetone, nail polish remover, heated chemicals, mothballs, and any other chemicals labeled or known to be dangerous	• Wear proper protective equipment such as goggles, apron, and gloves when using chemicals. • Ensure proper room ventilation or use a fume hood when using materials that produce fumes. • NEVER smell fumes directly. • NEVER taste or eat any material in the laboratory.	• If contact occurs, immediately flush affected area with water and notify your teacher. • If a spill occurs, leave the area immediately and notify your teacher.
FLAMMABLE	unexpected fire due to liquids or gases that ignite easily such as rubbing alcohol	• Avoid open flames, sparks, or heat when flammable liquids are present.	• If a fire occurs, leave the area immediately and notify your teacher.
OPEN FLAME	burns or fire due to open flame from matches, Bunsen burners, or burning materials	• Tie back loose hair and clothing. • Keep flame away from all materials. • Follow teacher instructions when lighting and extinguishing flames. • Use proper protection, such as hot mitts or tongs, when handling hot objects.	• If a fire occurs, leave the area immediately and notify your teacher.
ANIMAL SAFETY	injury to or from laboratory animals	• Wear proper protective equipment such as gloves, apron, and goggles when working with animals. • Wash hands after handling animals.	• If injury occurs, notify your teacher immediately.
BIOLOGICAL	infection or adverse reaction due to contact with organisms such as bacteria, fungi, and biological materials such as blood, animal or plant materials	• Wear proper protective equipment such as gloves, goggles, and apron when working with biological materials. • Avoid skin contact with an organism or any part of the organism. • Wash hands after handling organisms.	• If contact occurs, wash the affected area and notify your teacher immediately.
FUME	breathing difficulties from inhalation of fumes from substances such as ammonia, acetone, nail polish remover, heated chemicals, and mothballs	• Wear goggles, apron, and gloves. • Ensure proper room ventilation or use a fume hood when using substances that produce fumes. • NEVER smell fumes directly.	• If a spill occurs, leave area and notify your teacher immediately.
IRRITANT	irritation of skin, mucous membranes, or respiratory tract due to materials such as acids, bases, bleach, pollen, mothballs, steel wool, and potassium permanganate	• Wear goggles, apron, and gloves. • Wear a dust mask to protect against fine particles.	• If skin contact occurs, immediately flush the affected area with water and notify your teacher.
RADIOACTIVE	excessive exposure from alpha, beta, and gamma particles	• Remove gloves and wash hands with soap and water before removing remainder of protective equipment.	• If cracks or holes are found in the container, notify your teacher immediately.

Your online portal to everything you need

connectED.mcgraw-hill.com

Look for these icons to access
exciting digital resources

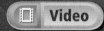 Video

 Audio

Review

? Inquiry

WebQuest

✓ Assessment

((○ Concepts in Motion

ENERGY AND MATTER

iSCIENCE

Glencoe

Bubbles

The iridescent colors of these soap bubbles result from a property called interference. Light waves reflect off both outside and inside surfaces of bubbles. When this happens, the waves interfere with each other and you see different colors. The thickness of the soap film that forms a bubble also affects interference.

The McGraw-Hill Companies

 Education

Send all inquiries to:
McGraw-Hill Education
8787 Orion Place
Columbus, OH 43240-4027

ISBN: 978-0-07-888020-9
MHID: 0-07-888020-3

Printed in the United States of America.

1 2 3 4 5 6 7 8 9 10 DOW 15 14 13 12 11

Authors

American Museum of Natural History
New York, NY

Michelle Anderson, MS
Lecturer
The Ohio State University
Columbus, OH

Juli Berwald, PhD
Science Writer
Austin, TX

John F. Bolzan, PhD
Science Writer
Columbus, OH

Rachel Clark, MS
Science Writer
Moscow, ID

Patricia Craig, MS
Science Writer
Bozeman, MT

Randall Frost, PhD
Science Writer
Pleasanton, CA

Lisa S. Gardiner, PhD
Science Writer
Denver, CO

Jennifer Gonya, PhD
The Ohio State University
Columbus, OH

Mary Ann Grobbel, MD
Science Writer
Grand Rapids, MI

Whitney Crispen Hagins, MA, MAT
Biology Teacher
Lexington High School
Lexington, MA

Carole Holmberg, BS
Planetarium Director
Calusa Nature Center and
Planetarium, Inc.
Fort Myers, FL

Tina C. Hopper
Science Writer
Rockwall, TX

Jonathan D. W. Kahl, PhD
Professor of Atmospheric Science
University of Wisconsin-
Milwaukee
Milwaukee, WI

Nanette Kalis
Science Writer
Athens, OH

S. Page Keeley, MEd
Maine Mathematics and
Science Alliance
Augusta, ME

Cindy Klevickis, PhD
Professor of Integrated Science
and Technology
James Madison University
Harrisonburg, VA

Kimberly Fekany Lee, PhD
Science Writer
La Grange, IL

Michael Manga, PhD
Professor
University of California, Berkeley
Berkeley, CA

Devi Ried Mathieu
Science Writer
Sebastopol, CA

Elizabeth A. Nagy-Shadman, PhD
Geology Professor
Pasadena City College
Pasadena, CA

William D. Rogers, DA
Professor of Biology
Ball State University
Muncie, IN

Donna L. Ross, PhD
Associate Professor
San Diego State University
San Diego, CA

Marion B. Sewer, PhD
Assistant Professor
School of Biology
Georgia Institute of Technology
Atlanta, GA

Julia Meyer Sheets, PhD
Lecturer
School of Earth Sciences
The Ohio State University
Columbus, OH

Michael J. Singer, PhD
Professor of Soil Science
Department of Land, Air and
Water Resources
University of California
Davis, CA

Karen S. Sottosanti, MA
Science Writer
Pickerington, Ohio

Paul K. Strode, PhD
I.B. Biology Teacher
Fairview High School
Boulder, CO

Jan M. Vermilye, PhD
Research Geologist
Seismo-Tectonic Reservoir
Monitoring (STRM)
Boulder, CO

Judith A. Yero, MA
Director
Teacher's Mind Resources
Hamilton, MT

Dinah Zike, MEd
Author, Consultant,
Inventor of Foldables
Dinah Zike Academy;
Dinah-Might Adventures, LP
San Antonio, TX

Margaret Zorn, MS
Science Writer
Yorktown, VA

Consulting Authors

Alton L. Biggs
Biggs Educational Consulting
Commerce, TX

Ralph M. Feather, Jr., PhD
Assistant Professor
Department of Educational
Studies and Secondary
Education
Bloomsburg University
Bloomsburg, PA

Douglas Fisher, PhD
Professor of Teacher Education
San Diego State University
San Diego, CA

Edward P. Ortleb
Science/Safety Consultant
St. Louis, MO

Series Consultants

Science

Solomon Bililign, PhD
Professor
Department of Physics
North Carolina Agricultural
and Technical State University
Greensboro, NC

John Choinski
Professor
Department of Biology
University of Central Arkansas
Conway, AR

Anastasia Chopelas, PhD
Research Professor
Department of Earth and
Space Sciences
UCLA
Los Angeles, CA

David T. Crowther, PhD
Professor of Science Education
University of Nevada, Reno
Reno, NV

A. John Gatz
Professor of Zoology
Ohio Wesleyan University
Delaware, OH

Sarah Gille, PhD
Professor
University of California
San Diego
La Jolla, CA

David G. Haase, PhD
Professor of Physics
North Carolina State
University
Raleigh, NC

Janet S. Herman, PhD
Professor
Department of Environmental
Sciences
University of Virginia
Charlottesville, VA

David T. Ho, PhD
Associate Professor
Department of Oceanography
University of Hawaii
Honolulu, HI

Ruth Howes, PhD
Professor of Physics
Marquette University
Milwaukee, WI

**Jose Miguel Hurtado, Jr.,
PhD**
Associate Professor
Department of Geological
Sciences
University of Texas at El Paso
El Paso, TX

Monika Kress, PhD
Assistant Professor
San Jose State University
San Jose, CA

Mark E. Lee, PhD
Associate Chair & Assistant
Professor
Department of Biology
Spelman College
Atlanta, GA

Linda Lundgren
Science writer
Lakewood, CO

Carolyn Elliott
Iredell-Statesville Schools
Statesville, NC

Christine M. Jacobs
Ranger Middle School
Murphy, NC

Jason O. L. Johnson
Thurmont Middle School
Thurmont, MD

Felecia Joiner
Stony Point Ninth Grade
Center
Round Rock, TX

Joseph L. Kowalski, MS
Lamar Academy
McAllen, TX

Brian McClain
Amos P. Godby High School
Tallahassee, FL

Von W. Mosser
Thurmont Middle School
Thurmont, MD

Ashlea Peterson
Heritage Intermediate Grade
Center
Coweta, OK

Nicole Lenihan Rhoades
Walkersville Middle School
Walkersvillle, MD

Maria A. Rozenberg
Indian Ridge Middle School
Davie, FL

Barb Seymour
Westridge Middle School
Overland Park, KS

Ginger Shirley
Our Lady of Providence
Junior-Senior High School
Clarksville, IN

Curtis Smith
Elmwood Middle School
Rogers, AR

Sheila Smith
Jackson Public School
Jackson, MS

Sabra Soileau
Moss Bluff Middle School
Lake Charles, LA

Tony Spoores
Switzerland County Middle
School
Vevay, IN

Nancy A. Stearns
Switzerland County Middle
School
Vevay, IN

Kari Vogel
Princeton Middle School
Princeton, MN

Alison Welch
Wm. D. Slider Middle School
El Paso, TX

Linda Workman
Parkway Northeast Middle
School
Creve Coeur, MO

Teacher Advisory Board

The Teacher Advisory Board gave the authors, editorial staff, and design team feedback on the content and design of the Student Edition. They provided valuable input in the development of *Glencoe ⓘScience.*

Frances J. Baldridge
Department Chair
Ferguson Middle School
Beavercreek, OH

Jane E. M. Buckingham
Teacher
Crispus Attucks Medical
Magnet High School
Indianapolis, IN

Elizabeth Falls
Teacher
Blalack Middle School
Carrollton, TX

Nelson Farrier
Teacher
Hamlin Middle School
Springfield, OR

Michelle R. Foster
Department Chair
Wayland Union
Middle School
Wayland, MI

Rebecca Goodell
Teacher
Reedy Creek Middle School
Cary, NC

Mary Gromko
Science Supervisor K–12
Colorado Springs District 11
Colorado Springs, CO

Randy Mousley
Department Chair
Dean Ray Stucky
Middle School
Wichita, KS

David Rodriguez
Teacher
Swift Creek Middle School
Tallahassee, FL

Derek Shook
Teacher
Floyd Middle Magnet School
Montgomery, AL

Karen Stratton
Science Coordinator
Lexington School District One
Lexington, SC

Stephanie Wood
Science Curriculum Specialist,
K–12
Granite School District
Salt Lake City, UT

Online Guide

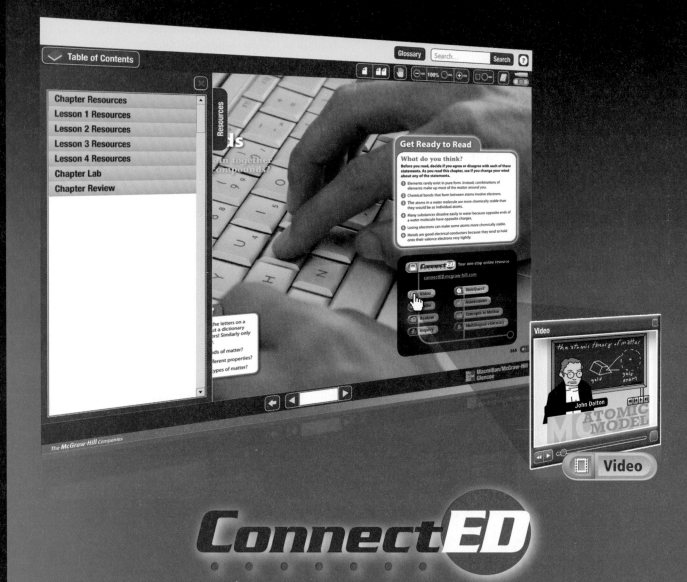

ConnectED

▷ **Your Digital Science Portal**

 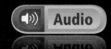   

See the science in real life through these exciting videos.

Click the link and you can listen to the text while you follow along.

Try these interactive tools to help you review the lesson concepts.

Explore concepts through hands-on and virtual labs.

These web-based challenges relate the concepts you're learning about to the latest news and research.

The icons in your online student edition link you to interactive learning opportunities. Browse your online student book to find more.

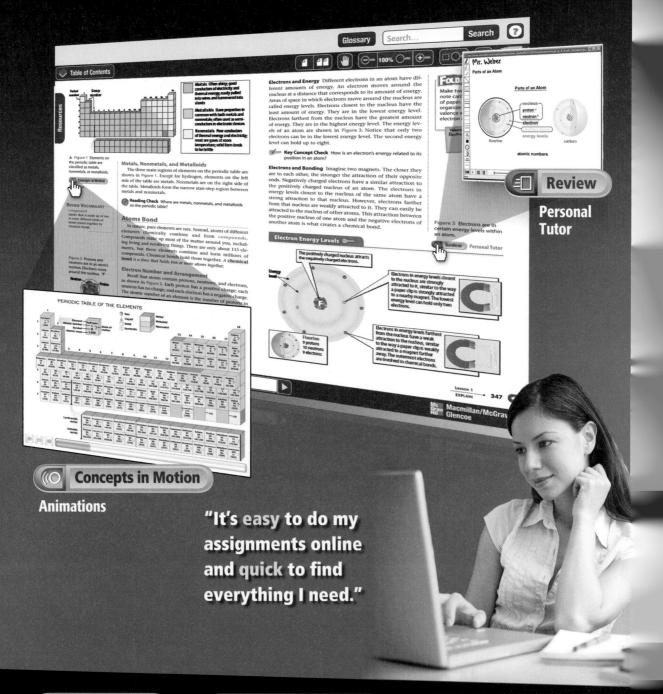

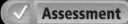
Concepts in Motion
Animations

"It's easy to do my assignments online and quick to find everything I need."

✓ **Assessment**
Check how well

((○)) **Concepts in Motion**
The textbook comes alive

g **Multilingual eGlossary**
Read key vocabulary in 13 languages

Treasure Hunt

Your science book has many features that will aid you in your learning. Some of these features are listed below. You can use the activity at the right to help you find these and other special features in the book.

- **THE BIG IDEA** can be found at the start of each chapter.

- The Reading Guide at the start of each lesson lists 🔑 **Key Concepts,** vocabulary terms, and online supplements to the content.

- 💻 **ConnectED** icons direct you to online resources such as animations, personal tutors, math practices, and quizzes.

- **Inquiry** Labs and Skill Practices are in each chapter.

- Your **FOLDABLES** help organize your notes.

1 What four margin items can help you build your vocabulary?

2 On what page does the glossary begin? What glossary is online?

3 In which Student Resource at the back of your book can you find a listing of Laboratory Safety Symbols?

4 Suppose you want to find a list of all the Launch Labs, MiniLabs, Skill Practices, and Labs, where do you look?

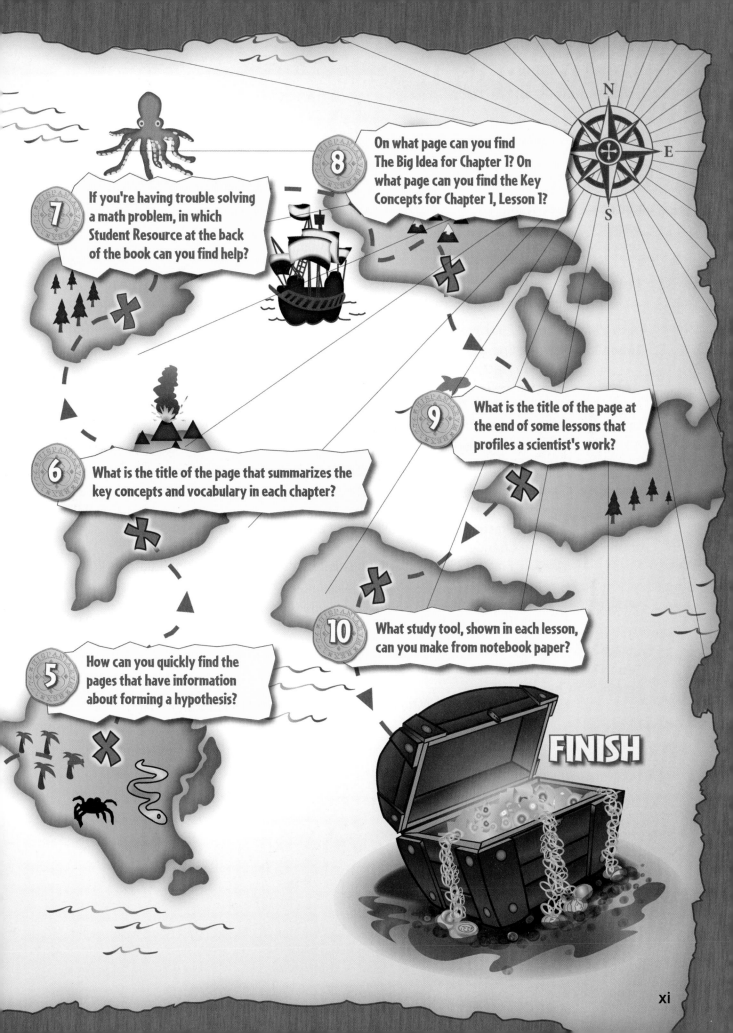

Table of Contents

Unit 2   **Energy and Matter** **154**

Chapter 5   **Energy and Energy Resources** **158**
Lesson 1   Forms of Energy 160
Lesson 2   Energy Transformations 168
     (Inquiry) **Skill Practice** Can you identify energy transformations? 175
Lesson 3   Energy Resources 176
     (Inquiry) **Lab** Pinwheel Power.. 186

Chapter 6   **Thermal Energy** **194**
Lesson 1   Thermal Energy, Temperature, and Heat 196
     (Inquiry) **Skill Practice** How do different materials affect thermal energy
       transfer? 203
Lesson 2   Thermal Energy Transfers.................................... 204
Lesson 3   Using Thermal Energy.. 214
     (Inquiry) **Lab** Design an Insulated Container 220

Chapter 7   **Foundations of Chemistry** **228**
Lesson 1   Classifying Matter.. 230
Lesson 2   Physical Properties... 239
     (Inquiry) **Skill Practice** How can following a procedure help you solve a crime? ..247
Lesson 3   Physical Changes 248
     (Inquiry) **Skill Practice** How can known substances help you identify unknown
       substances?... 254
Lesson 4   Chemical Properties and Changes 255
     (Inquiry) **Lab** Design an Experiment to Solve a Crime........................ 262

Chapter 8   **States of Matter**.. **270**
Lesson 1   Solids, Liquids, and Gases 272
Lesson 2   Changes in State... 281
     (Inquiry) **Lab** Skill Practice How does dissolving substances in water change
       its freezing point? 290
Lesson 3   The Behavior of Gases.. 291
     (Inquiry) **Lab** Design an Experiment to Collect Data 298

TABLE OF CONTENTS

Table of Contents

Student Resources

Science Skill Handbook **SR-2**

Scientific Methods SR-2

Safety Symbols SR-11

Safety in the Science Laboratory SR-12

Math Skill Handbook **SR-14**

Math Review SR-14

Science Application SR-24

Foldables Handbook **SR-29**

Reference Handbook **SR-40**

Periodic Table of the Elements SR-40

Glossary **G-2**

Index **I-2**

Credits **C-2**

Inquiry

Inquiry Launch Labs

5-1 Can you make a change in matter?.. 161

5-2 Is energy lost when it changes form?... 169

5-3 How are energy resources different? 177

6-1 How can you describe temperature? 197

6-2 How hot is it?.. 205

6-3 How can you transform energy? 215

7-1 How do you classify matter? 231

7-2 Can you follow the clues?... 240

7-3 Where did it go?... 249

7-4 What can colors tell you? 256

8-1 How can you see particles in matter?... 273

8-2 Do liquid particles move?... 282

8-3 Are volume and pressure of a gas related? 292

Inquiry MiniLabs

5-1 Can a moving object do work?.. 164

5-2 How does energy change form? 173

5-3 What energy resources provide our electric energy?................................... 183

6-1 How do temperature scales compare? 201

6-2 How does adding thermal energy affect a wire?....................................... 209

6-3 Can thermal energy be used to do work?... 217

7-1 How can you model an atom? 232

7-2 Can the weight of an object change? 242

7-3 Can you make ice without a freezer? 251

7-4 Can you spot the clues for chemical change?... 258

8-1 How can you make bubble films? 277

8-2 How can you make a water thermometer? 288

8-3 How does temperature affect the volume? 295

Inquiry Skill Practice

5-2 Can you identify energy transformations? 175

6-1 How do different materials affect thermal energy transfer? 203

7-2 How can following a procedure help you solve a crime? 247

7-3 How can known substances help you identify unknown substances? 254

8-2 How does dissolving substances in water change its freezing point? 290

Inquiry Labs

5-3 Pinwheel Power 186

6-3 Design an Insulated Container 220

7-4 Design an Experiment to Solve a Crime 262

8-3 Design an Experiment to Collect Data 298

Features

GREEN SCIENCE

5-1 Fossil Fuels and Rising CO_2 Levels.. 167

HOW IT WORKS

7-1 U.S. Mint... 238

8-1 Freeze-Drying Foods 280

SCIENCE & SOCIETY

6-2 Insulating the Home 213

ENERGY AND MATTER

1875   **1900**   **1925**

1895
The first X-ray photograph is taken by Wilhelm Konrad Roentgen of his wife's hand. It is now possible to look inside the human body without surgical intervention.

1898
Chemist Marie Curie and her husband Pierre discover radioactivity. They are later awarded the Nobel Prize in Physics for their discovery.

1917
Ernest Rutherford, the "father of nuclear physics," is the first to split atoms.

1934
Nuclear fission is first achieved experimentally in Rome by Enrico Fermi when his team bombards uranium with neutrons.

1939
The Manhattan Project, a code name for a research program to develop the first atomic bomb, begins. The project is directed by American physicist J. Robert Oppenheimer.

1950

1975

2000

1945
American-led atomic
bomb attacks on the
Japanese cities of
Hiroshima and
Nagasaki bring
about the end of
World War II.

1954
Obninsk Nuclear Power Plant,
located in the former USSR,
begins operating as the
world's first nuclear power
plant to generate electricity
for a power grid. It produces
around 5 megawatts of
electric power.

2007
Fourteen percent
of the world's
electricity now
comes from
nuclear power.

? Inquiry

**Visit ConnectED for
this unit's
STEM activity.**

Technology

Scientists use technology to develop materials with desirable properties. **Technology** is the practical use of scientific knowledge, especially for industrial or commercial use. In the late 1800s, scientists developed the first plastic material, called celluloid, from cotton. Celluloid quickly gained popularity for use as photographic film. In the 20th century, scientists developed other plastic materials, such as polystyrene, rayon, and nylon. These new materials were inexpensive, durable, lightweight, and could be molded into any shape.

New technologies can come with problems. For example, many plastics are made from petroleum and contain harmful chemicals. The high pressures and temperatures needed to produce plastics require large amounts of energy. Bacteria and fungi that easily break down natural materials do not easily decompose plastics. Often, plastics accumulate in landfills where they can remain for hundreds, or even thousands, of years, as shown in **Figure 1.**

Figure 1 Nature cannot easily recycle many human-made materials. Much of our trash remains in landfills for years. Scientists are developing materials that degrade quickly. This will help decrease the amount of pollution.

Types of Materials

Figure 2 Some organisms produce materials with properties that are useful to people. Scientists are trying to replicate these materials for new technologies.

◀ Most human-made adhesives attach to some surfaces, but not others. Mussels, which are similar to clams, produce a "superglue" that is stronger than anything people can make. It also works on any surface, wet or dry. Chemists are trying to develop a technology that will replicate the mussel glue. This glue would provide solutions to difficult problems: Ships could be repaired under water. The glue also would work on teeth and could be used to set broken bones.

Abalone and other mollusks construct a protective shell from proteins and seawater. The material is four times stronger than human-made metal alloys and ceramics. Using technology, scientists are working to duplicate this material. They hope to use the new product in many ways, including hip and elbow replacements. Automakers could use these strong, lightweight materials for automobile body panels. ▶

Consider the Possibilities!

Chemists are looking to nature for ideas for new materials. For example, some sea sponges have skeletons that beam light deep inside the animal, similar to the way fiber-optic cables work. A bacterium from a snail-like nudibranch contains compounds that stop other sea creatures from growing on the nudibranch's back. These compounds could be used in paints to stop creatures from forming a harmful crust on submerged parts of boats and docks. Chrysanthemum flowers produce a product that keeps ticks and mosquitoes away. **Figure 2** includes other organisms that produce materials with remarkable properties.

Chemists and biologists are teaming up to understand, and hopefully replicate, the processes that organisms use to survive. Hopefully, these processes can lead to technologies and materials with unique properties that are helpful to people.

Inquiry MiniLab
15 minutes

How would you use it?

How would you use an adhesive that could stick to any surface? Invent a new purpose for mussel "superglue"!

1. Work with a partner to develop three tasks that could be accomplished using products produced by an organism.

2. Select one of your ideas, and develop it into an invention. Draw pictures of your new invention and explain how it works.

3. Write an advertisement for your invention including a description of the role of the material from nature used in your product.

Analyze and Conclude

1. **Explain** What task or problem did your invention solve?

2. **Infer** How did a material from an organism help you develop your invention?

A British company has developed bacteria that produce large amounts of hydrogen gas when fed a diet of sugar. Chemists are working to produce tanks of these microorganisms that produce enough hydrogen to replace other fuels used to heat homes. Bacteria may become the power plants of the future.

Under a microscope, the horn of a rhinoceros looks much like the material used to make the wings of a Stealth aircraft. However, the rhino horn is self-healing. Picture a car with technologically advanced fenders similar to the horn of a rhinoceros; such a car could repair itself if it were in a fender-bender! ▶

Spider silk begins as a liquid inside the spider's body. When ejected through openings, called spinnerets, it becomes similar to a plastic thread. However, its properties include strength five times greater than steel, stretchability greater than nylon, and toughness better than the material in bulletproof vests! Chemists are using technology to make a synthetic spider silk. They hope to someday use the material for cables strong enough to support a bridge or as reinforcing fibers in aircraft bodies.

Chapter 5

Energy and Energy Resources

THE BIG IDEA What is energy and what are energy resources?

Which objects have energy?

If your answer is "everything in the photo," you are right. All objects contain energy. Some objects contain more energy than other objects. The Sun contains so much energy it is considered an energy resource.

• Where do you think the energy comes from that powers the cars?

• Do you think the energy in the Sun and the energy in the green plants are related?

• What do the terms *energy* and *energy resources* mean to you?

Get Ready to Read

What do you think?

Before you read, decide if you agree or disagree with each of these statements. As you read this chapter, see if you change your mind about any of the statements.

1 A fast-moving baseball has more kinetic energy than a slow-moving baseball.

2 A book sitting on a shelf has no energy.

3 Energy can change from one form to another.

4 If you toss a baton straight up, total energy decreases as the baton rises.

5 Nuclear power plants release many dangerous pollutants into the air as they transform nuclear energy into electric energy.

6 Thermal energy from within Earth can be transformed into electric energy at a power plant.

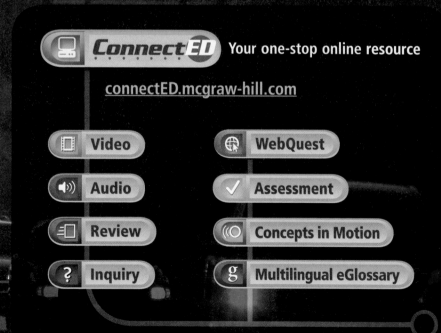

ConnectED Your one-stop online resource

connectED.mcgraw-hill.com

- Video
- Audio
- Review
- Inquiry
- WebQuest
- Assessment
- Concepts in Motion
- Multilingual eGlossary

Forms of Energy

Reading Guide

Key Concepts 🔑
ESSENTIAL QUESTIONS

- What is energy?
- What are potential and kinetic energy?
- How is energy related to work?
- What are different forms of energy?

Vocabulary

energy p. 161

kinetic energy p. 162

potential energy p. 162

work p. 164

mechanical energy p. 165

sound energy p. 165

thermal energy p. 165

electric energy p. 165

radiant energy p. 165

nuclear energy p. 165

 Multilingual eGlossary

 Video **BrainPOP®**

Inquiry Why is this cat glowing?

A camera that detects temperature made this image. Dark colors represent cooler temperatures and light colors represent warmer temperatures. Temperatures are cooler where the cat's body emits less radiant energy and warmer where the cat's body emits more radiant energy.

Inquiry Launch Lab

20 minutes

Can you make a change in matter?

You observe many things changing. Birds change their positions when they fly. Bubbles form in boiling water. The filament in a lightbulb glows when you turn on a light. How can you cause a change in matter?

1. Read and complete the lab safety form.
2. Half-fill a **foam cup** with **sand.** Place the bulb of a **thermometer** about halfway into the sand. *Do not stir.* Record the temperature in your Science Journal.
3. Remove the thermometer and place a **lid** on the cup. Hold down the lid and shake the cup vigorously for 10 minutes.
4. Remove the lid. Measure and record the temperature of the sand.

Think About This

1. What change did you observe in the sand?
2. **Predict** how you could change your results.
3. **Key Concept** What do you think caused the change?

What is energy?

It might be exciting to watch a fireworks display like the one shown in **Figure 1.** Over and over, you hear the crack of explosions and see bursts of colors in the night sky. Fireworks release energy when they explode. **Energy** *is the ability to cause change.* The energy in the fireworks causes the changes you see as bursting flashes of light and hear as loud booms.

Energy also causes other changes. The plant in **Figure 1** uses the energy from the Sun and makes food that it uses for growth and other processes. Energy can cause changes in the motions and positions of objects, such as the nail in **Figure 1.** Can you think of other ways energy might cause changes?

Key Concept Check What is energy?

WORD ORIGIN ············

energy
from Greek *energeia*, means "activity"

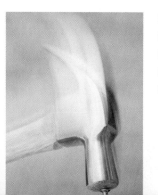

Figure 1 The explosion of the fireworks, the growth of the flower, and the motion of the hammer all involve energy.

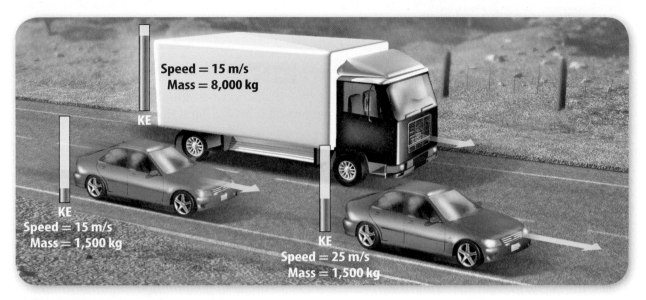

Speed = 15 m/s
Mass = 8,000 kg

KE

KE
Speed = 15 m/s
Mass = 1,500 kg

KE
Speed = 25 m/s
Mass = 1,500 kg

Figure 2 🔑 The kinetic energy (KE) of an object depends on its speed and its mass. The vertical bars show the kinetic energy of each vehicle.

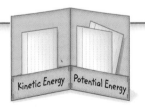
Kinetic Energy—Energy of Motion

Have you ever been to a bowling alley? When you rolled the ball and it hit the pins, a change occurred—the pins fell over. This change occurred because the ball had a form of energy called kinetic (kuh NEH tik) energy. **Kinetic energy** *is energy due to motion.* All moving objects have kinetic energy.

Kinetic Energy and Speed

An object's kinetic energy depends on its speed. The faster an object moves, the more kinetic energy it has. For example, the blue car has more kinetic energy than the green car in **Figure 2** because the blue car is moving faster.

Kinetic Energy and Mass

A moving object's kinetic energy also depends on its mass. If two objects move at the same speed, the object with more mass has more kinetic energy. For example, the truck and the green car in **Figure 2** are moving at the same speed, but the truck has more kinetic energy because it has more mass.

 Key Concept Check What is kinetic energy?

Potential Energy—Stored Energy

Energy can be present even if objects are not moving. If you hold a ball in your hand and then let it go, the gravitational interaction between the ball and Earth causes a change to occur. Before you dropped the ball, it had a form of energy called potential (puh TEN chul) energy. **Potential energy** *is stored energy due to the interactions between objects or particles.* Gravitational potential energy, elastic potential energy, and chemical potential energy are all forms of potential energy.

Gravitational Potential Energy

Even when you are just holding a book, energy is stored between the book and Earth. This type of energy is called gravitational potential energy. The girl in **Figure 3** increases the gravitational potential energy between her backpack and Earth by lifting the backpack.

The gravitational potential energy stored between an object and Earth depends on the object's weight and height. Dropping a bowling ball from a height of 1 m causes a greater change than dropping a tennis ball from 1 m. Similarly, dropping a bowling ball from 3 m causes a greater change than dropping the same bowling ball from only 1 m.

 Reading Check What factors determine the gravitational potential energy stored between an object and Earth?

Elastic Potential Energy

When you stretch a rubber band, like the one in **Figure 3,** you are storing another form of potential energy called elastic (ih LAS tik) potential energy. Elastic potential energy is energy stored in objects that are compressed or stretched, such as springs and rubber bands. When you release the end of a stretched rubber band, the stored elastic potential energy is transformed into kinetic energy.

Chemical Potential Energy

Food, gasoline, and other substances are made of atoms joined together by chemical bonds. Chemical potential energy is energy stored in the chemical bonds between atoms, as shown in **Figure 3.** Chemical potential energy is released when chemical reactions occur. Your body uses the chemical potential energy in foods for all its activities. People also use the chemical potential energy in gasoline to drive cars and buses.

 Key Concept Check In what way are all forms of potential energy the same?

Potential Energy 🔑

Figure 3 There are different forms of potential energy.

Gravitational Potential Energy
Gravitational potential energy increases when the girl lifts her backpack.

Elastic Potential Energy
The rubber band's elastic potential energy increases when you stretch the rubber band.

Chemical Potential Energy
Foods and other substances, including glucose, have chemical potential energy stored in the bonds between atoms.

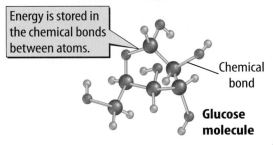

Energy is stored in the chemical bonds between atoms.

Chemical bond

Glucose molecule

Figure 4 The girl does work on the box as she lifts it. The work she does transfers energy to the box. The colored bars show the work that the girl does (W) and the box's potential energy (PE).

Energy and Work

You can transfer energy by doing work. **Work** *is the transfer of energy that occurs when a force is applied over a distance.* For example, the girl does work on the box in **Figure 4.** As the girl lifts the box onto the shelf, she transfers energy from herself to the gravitational interaction between the box and Earth.

Work depends on both force and distance. You only do work on an object if that object moves. Imagine that the girl in **Figure 4** tries to lift the box but cannot actually lift it off the floor. Then she does no work on the box and transfers no energy.

 Key Concept Check How is energy related to work?

An object that has energy also can do work. For example, when a bowling ball collides with a bowling pin, the bowling ball does work on the pin. Some of the ball's kinetic energy is transferred to the bowling pin. Because of this connection between energy and work, energy is sometimes described as the ability to do work.

Other Forms of Energy

Some other forms of energy are shown in **Table 1.** All energy can be measured in joules (J). A softball dropped from a height of about 0.5 m has about 1 J of kinetic energy just before it hits the floor.

Inquiry) MiniLab

20 minutes

Can a moving object do work?

Is work done when a moving object hits another object?

1 Read and complete a lab safety form.

2 **Tape** one end of a **30-cm grooved ruler** to the edge of a stack of **books** about 8 cm high. Put the lower end of the ruler in a **paper cup** laying on its side.

3 Release a **marble** in the groove at the top end of the ruler.

4 Record your observations in your Science Journal.

Analyze and Conclude

1. **Compare** the kinetic energy of the marble just before and after it hit the cup.

2. 🔑 **Key Concept** Is work being done on the cup? Explain your answer.

Table 1 Forms of Energy 🔑

Mechanical Energy

The total energy of an object or group of objects due to large-scale motions and interactions is called **Mechanical energy**. For example, the mechanical energy of a basketball increases when a player shoots the basketball. However, the mechanical energy of a pot of water does not increase when you heat the water.

Sound Energy

When you pluck a guitar string, the string vibrates and produces sound. *The energy that sound carries is* **sound energy.** Vibrating objects emit sound energy. However, sound energy cannot travel through a vacuum such as the space between Earth and the Sun.

Thermal Energy

All objects and materials are made of particles that are always moving. Because these particles move, they have energy. **Thermal energy** *is energy due to the motion of particles that make up an object.* Thermal energy moves from warmer objects to colder objects. When you heat objects, you transfer thermal energy to those objects from their surroundings.

Electric Energy

An electric fan uses another form of energy—electric energy. When you turn on a fan, there is an electric current through the fan's motor. **Electric energy** *is the energy that an electric current carries.* Electric appliances, such as fans and dishwashers, change electric energy into other forms of energy.

Radiant Energy—Light Energy

The Sun gives off energy that travels to Earth as electromagnetic waves. Unlike sound waves, electromagnetic waves can travel through a vacuum. Light waves, microwaves, and radio waves are all electromagnetic waves. *The energy that electromagnetic waves carry is* **radiant energy.** Sometimes radiant energy is called light energy.

Nuclear Energy

At the center of every atom is a nucleus. **Nuclear energy** *is energy that is stored in the nucleus of an atom.* In the Sun, nuclear energy is released when nuclei join together. In a nuclear power plant, nuclear energy is released when the nuclei of uranium atoms are split apart.

 Key Concept Check Describe three forms of energy.

Lesson 1 Review

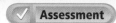
Visual Summary

Energy is the ability to cause change.

The gravitational potential energy between an object and Earth increases when you lift the object.

You do work on an object when you apply a force to that object over a distance.

FOLDABLES

Use your lesson Foldable to review the lesson. Save your Foldable for the project at the end of the chapter.

What do you think NOW?

You first read the statements below at the beginning of the chapter.

1. A fast-moving baseball has more kinetic energy than a slow-moving baseball.

2. A book sitting on a shelf has no energy.

Did you change your mind about whether you agree or disagree with the statements? Rewrite any false statements to make them true.

Use Vocabulary

1 **Distinguish** between kinetic energy and potential energy.

2 **Write** a definition of work.

Understand Key Concepts

3 Which type of energy increases when you compress a spring?
- A. elastic potential energy
- B. kinetic energy
- C. radiant energy
- D. sound energy

4 **Infer** How could you increase the gravitational potential energy between yourself and Earth?

5 **Infer** how a bicycle's kinetic energy changes when that bicycle slows down.

6 **Compare and contrast** radiant energy and sound energy.

Interpret Graphics

7 **Identify** Copy and fill in the graphic organizer below to identify three types of potential energy.

8 **Describe** where chemical potential energy is stored in the molecule shown below.

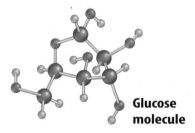

Glucose molecule

Critical Thinking

9 **Analyze** Will pushing on a car always change the car's mechanical energy? What must happen for the car's kinetic energy to increase?

Fossil Fuels and Rising CO$_2$ Levels

Investigate the link between energy use and carbon dioxide in the atmosphere.

You use energy every day—when you ride in a car or on a bus, turn on a television, and even when you send an e-mail.

Much of the energy that produces electric current, heats and cools buildings, and powers engines, comes from burning fossil fuels—coal, oil, and natural gas. When fossil fuels burn, the carbon in them combines with oxygen in the atmosphere and forms carbon dioxide gas (CO$_2$). Carbon dioxide is one of the greenhouse gases. In the atmosphere, greenhouse gases absorb energy. This causes the atmosphere and Earth's surface to become warmer. Greenhouse gases make Earth warm enough to support life. Without greenhouse gases, Earth's surface would be frozen.

However, over the past 150 years, the amount of CO$_2$ in the atmosphere has increased faster than at any time in the past 800,000 years. Most of this increase is the result of burning fossil fuels. This additional carbon dioxide will cause average global temperatures to increase. As temperatures increase, weather patterns worldwide could change. More storms and heavier rainfall could occur in some areas, while other regions could become drier. Increased temperatures also will cause more of the polar ice sheets to melts, causing sea levels to rise. Higher sea levels will cause more flooding in coastal areas.

Developing other energy sources, such as geothermal, solar, nuclear, wind, and hydroelectric power, would reduce the use of fossil fuels and slow the increase in atmospheric CO$_2$.

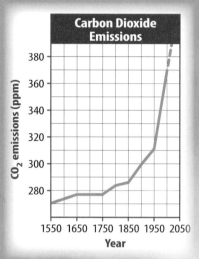

Carbon Dioxide Emissions

It's Your Turn

MAKE A LIST How can CO$_2$ emissions be reduced? Work with a partner. List five ways people in your home, school, or community could reduce their energy consumption. Combine your list with your classmates' lists to make a master list.

300 Years OF CARBON DIOXIDE

● **1712**
A new invention, the steam engine, is powered by burning coal that heats water to produce steam.

● **Early 1800s**
Coal-fired steam engines, able to pull heavy trains and power steamboats, transform transportation.

● **1882**
Companies make and sell electric energy from coal for everyday use. Electricity was used to power the first lightbulbs, which give off 20 times the light of a candle.

● **1908**
The first mass-produced automobiles are made available. By 1915, Ford was selling 500,000 cars a year. Oil becomes the fuel of choice for car engines.

● **Late 1900s**
Electric appliances transform the way we live, work, and communicate. Most electricity is generated by coal-burning power plants.

● **2007**
There are more than 800 million cars and light trucks on the world's roads.

Energy Transformations

Reading Guide

Key Concepts 🔑
ESSENTIAL QUESTIONS

- What is the law of conservation of energy?
- How does friction affect energy transformations?
- How are different types of energy used?

Vocabulary

law of conservation of energy p. 170

friction p. 171

g Multilingual eGlossary

Inquiry **What's that sound?**

Blocks of ice breaking off the front of this glacier can be bigger than a car. Imagine the loud rumble they make as they crash into the sea. But after the ice falls into the sea, it will gradually melt. All of these processes involve energy transformations—energy changing from one form to another.

Is energy lost when it changes form?

Energy can have different forms. What happens when energy changes from one form to another?

1. Read and complete the lab safety form.

2. Three students should sit in a circle. One student has 30 **buttons,** one has 30 **pennies,** and one has 30 **paper clips.**

3. Each student should exchange 10 items with the student to the right and 10 items with the student to the left.

4. Repeat step 3.

Think About This

1. If the buttons, pennies, and paper clips represented different forms of energy, what represented changes from one form of energy to another?

2. 🔑 **Key Concept** If each button, penny, and paper clip represented one unit of energy, did the total amount of energy increase, decrease, or stay the same? Explain your answer.

Changes Between Forms of Energy

It is the weekend and you are ready to make some popcorn in the microwave and watch a movie. Energy changes form when you make popcorn and watch TV. As shown in **Figure 5,** a microwave changes electric energy into **radiant** energy. Radiant energy changes into thermal energy in the popcorn kernels.

The changes from electric energy to radiant energy to thermal energy are called energy transformations. As you watch the movie, energy transformations also occur in the television. A television transforms electric energy into sound energy and radiant energy.

SCIENCE USE V. COMMON USE

radiant
Science Use transmitted by electromagnetic waves

Common Use bright and shining; glowing

Figure 5 Energy changes from one form to another when you use a microwave oven to make popcorn.

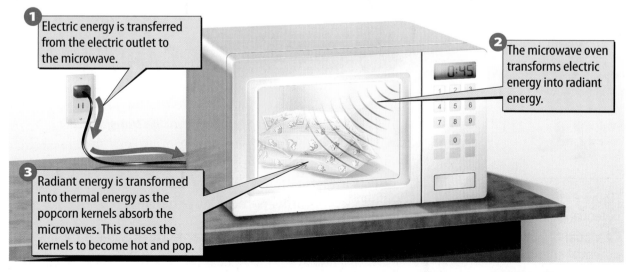

1. Electric energy is transferred from the electric outlet to the microwave.

2. The microwave oven transforms electric energy into radiant energy.

3. Radiant energy is transformed into thermal energy as the popcorn kernels absorb the microwaves. This causes the kernels to become hot and pop.

Conservation of Energy

Concepts in Motion Animation

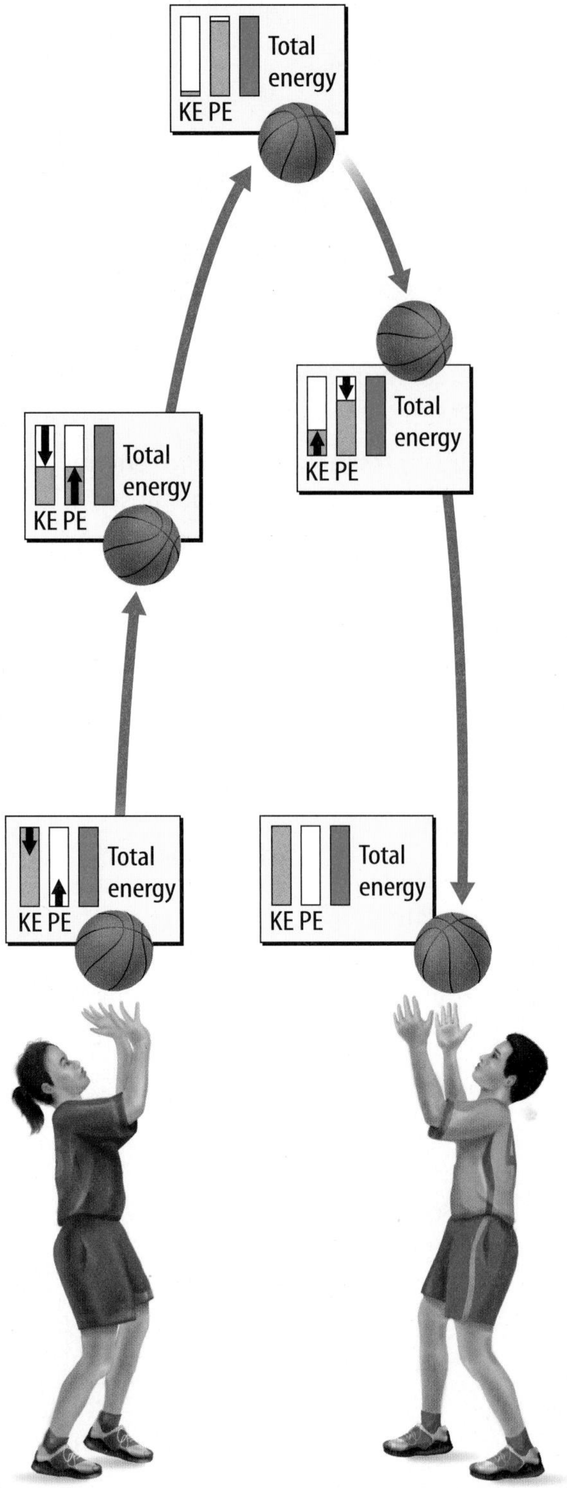

Figure 6 The ball's kinetic energy (KE) and potential energy (PE) changes as it moves.

Visual Check When is the gravitational potential energy the greatest?

Changes Between Kinetic and Potential Energy

Energy transformations also occur when you toss a ball upward, as shown in **Figure 6.** The ball slows down as it moves upward and then speeds up as it moves downward. The ball's speed and height change as energy changes from one form to another.

Kinetic Energy to Potential Energy

The ball is moving fastest and has the most kinetic energy as it leaves your hand, as shown in **Figure 6.** As the ball moves upward, its speed and kinetic energy decrease. However, the potential energy is increasing because the ball's height is increasing. Kinetic energy is changing into potential energy. At the ball's highest point, the gravitational potential energy is greatest, and the ball's kinetic energy is the least.

Potential Energy to Kinetic Energy

As the ball moves downward, potential energy decreases. At the same time, the ball's speed increases. Therefore, the ball's kinetic energy increases. Potential energy is transformed into kinetic energy. When the ball reaches the player's hand again, its kinetic energy is at the maximum value again.

Reading Check Why does the potential energy decrease as the ball falls?

The Law of Conservation of Energy

The total energy in the universe is the sum of all the different forms of energy everywhere. According to the **law of conservation of energy,** *energy can be transformed from one form into another or transferred from one region to another, but energy cannot be created or destroyed.* The total amount of energy in the universe does not change.

Key Concept Check What is the law of conservation of energy?

Friction and the Law of Conservation of Energy

Sometimes it may seem as if the law of conservation of energy is not accurate. Imagine riding a bicycle, as in **Figure 7.** The moving bicycle has mechanical energy. What happens to this mechanical energy when you apply the brakes and the bicycle stops?

When you apply the brakes, the bicycle's mechanical energy is not destroyed. Instead the bicycle's mechanical energy is transformed into thermal energy, as shown in **Figure 7.** The total amount of energy never changes. The additional thermal energy causes the brakes, the wheels, and the air around the bicycle to become slightly warmer.

Friction *is a force that resists the sliding of two surfaces that are touching.* Friction between the bicycle's brake pads and the moving wheels transforms mechanical energy into thermal energy.

 Key Concept Check Friction causes what energy transformation?

There is always some friction between any two surfaces that are rubbing against each other. As a result, some mechanical energy is always transformed into thermal energy when two surfaces rub against each other.

It is easier to pedal a bicycle if there is less friction between the bicycle's parts. With less friction, less of the bicycle's mechanical energy gets transformed into thermal energy. One way to reduce friction is to apply a lubricant, such as oil, grease, or graphite, to surfaces that rub against each other.

Friction and Thermal Energy

Review   Personal Tutor

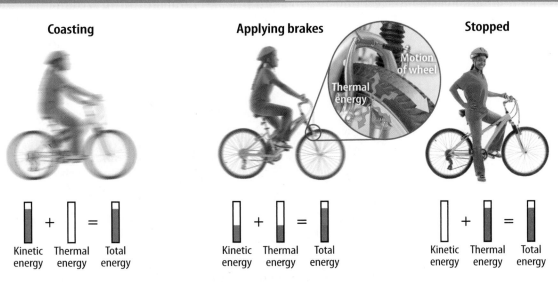

Figure 7 When the girl applies the brakes, friction between the bicycle's brake pads and its wheels transforms mechanical energy into thermal energy. As mechanical energy changes into thermal energy, the bicycle slows down. The total amount of energy does not change.

Using Energy

Every day you use different forms of energy to do different things. You might use the radiant energy from a lamp to light a room, or you might use the chemical energy stored in your body to run a race. When you use energy, you usually change it from one form into another. For example, the lamp changes electric energy into radiant energy and thermal energy.

Using Thermal Energy

All forms of energy can be transformed into thermal energy. People often use thermal energy to cook food or provide warmth. A gas stove transforms the chemical energy stored in natural gas into the thermal energy that cooks food. An electric space heater transforms the electric energy from a power plant into the thermal energy that warms a room. In a jet engine, burning fuel releases thermal energy that the engine transforms into mechanical energy.

Using Chemical Energy

During photosynthesis, a plant transforms the Sun's radiant energy into chemical energy that it stores in chemical compounds. Some of these compounds become food for other living things. Your body transforms the chemical energy from your food into the kinetic energy necessary for movement. Your body also transforms chemical energy into the thermal energy necessary to keep you warm.

Using Radiant Energy

The cell phone in **Figure 8** sends and receives radiant energy using microwaves. When you are listening to someone on a cell phone, that cell phone is transforming radiant energy into electrical energy and then into sound energy. When you are speaking into a cell phone, it is transforming sound energy into electric energy and then into radiant energy.

Cut three sheets of paper in half. Use the six half sheets to make a side-tab book with five tabs and a cover. Use your book to organize your notes on energy transformations.

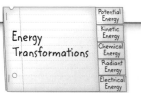

Figure 8 A cell phone changes sound energy into radiant energy when you speak.

Sound waves carry energy into the cell phone.

The cell phone converts the energy carried by sound waves into radiant energy that is carried away by microwaves.

Using Electric Energy

Many of the devices you might use every day, such as handheld video games, mp3 players, and hair dryers, use electric energy. Some devices, such as hair dryers, use electric energy from electrical power plants. Other appliances, such as handheld video games, transform the chemical energy stored in batteries into electric energy.

 Key Concept Check What happens to energy when it is used?

Waste Energy

When energy changes form, some thermal energy is always released. For example, a lightbulb converts some electric energy into radiant energy. However, the lightbulb also transforms some electric energy into thermal energy. This is what makes the lightbulb hot. Some of this thermal energy moves into the air and cannot be used.

Scientists often refer to thermal energy that cannot be used as waste energy. Whenever energy is used, some energy is transformed into useful energy and some is transformed into waste energy. For example, the drivers of the cars in **Figure 9** use the chemical energy in gasoline to make the cars move. However, most of that chemical energy ends up as waste energy—thermal energy that moves into the air.

 Reading Check What is waste energy?

 **MiniLab** 20 minutes

How does energy change form?

When an object falls, energy changes form. How can you compare energies of falling objects?

1. Read and complete a lab safety form.
2. Place a piece of **clay** about 10 cm wide and 3 cm thick on a **small paper plate.**
3. Drop a **marble** into the clay from a height of about 20 cm and measure the depth of the depression caused by the marble. Record the measurement in your Science Journal.
4. Repeat step 3 with a heavier marble.

Analyze and Conclude

1. **Infer** Which marble had more kinetic energy just before it hit the clay? Explain your answer.
2. **Key Concept** For which marble was the potential energy greater just before the marble fell? Explain your answer using the law of conservation of energy.

Figure 9 Cars transform most of the chemical energy in gasoline into waste energy.

Lesson 2 Review

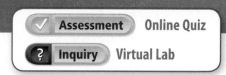
Visual Summary

Energy can change form, but follows the law of conservation of energy.

Friction can be reduced by applying a lubricant, such as oil or grease.

Different forms of energy, such as sound and radiant energy, are used when someone talks on a cell phone.

FOLDABLES

Use your lesson Foldable to review the lesson. Save your Foldable for the project at the end of the chapter.

What do you think NOW?

You first read the statements below at the beginning of the chapter.

3. Energy can change from one form to another.

4. If you toss a baton straight up, total energy decreases as the baton rises.

Did you change your mind about whether you agree or disagree with the statements? Rewrite any false statements to make them true.

Use Vocabulary

1 **Define** *friction* in your own words.

2 **Explain** the law of conservation of energy in your own words.

Understand Key Concepts

3 **Describe** the energy transformations that occur when a piece of wood burns.

4 **Identify** the energy transformation that takes place when you apply the brakes on a bicycle.

5 Which energy transformation occurs in a toaster?
 A. chemical to electrical
 B. electrical to thermal
 C. kinetic to chemical
 D. thermal to potential

Interpret Graphics

6 **Organize Information** Copy and fill in the graphic organizer below to show how kinetic and potential energy change when a ball is thrown straight up and then falls down.

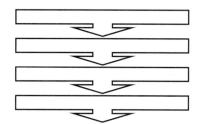

Critical Thinking

7 **Identify** Choose three electric appliances. For each appliance, identify the forms of energy that electric energy is transformed into.

8 **Judge** An advertisement states that a machine with moving parts will continue moving forever without having to add any energy. Can this advertisement be correct? Explain.

Can you identify energy transformations?

Energy cannot be created or destroyed. However, energy can be transformed from one type into another type, and energy can be transferred from one object to another object. In this lab, you will transform gravitational potential energy into kinetic energy as well as transfer energy from one object to another object.

Materials

string

paper clip

three large washers

meterstick

tape

small box

ruler

Safety

🥽

Learn It

Before you can draw valid conclusions from any scientific experiment, you must **analyze** the results of that experiment. This means you must look for patterns in the results.

Try It

1. Read and complete a lab safety form.

2. Use the photo below as a guide to make a pendulum. Hang one washer on the paper clip. Place the box so it will block the swinging pendulum. Mark the position of the box with tape.

3. Pull the pendulum back until the bottom of the washer is 15 cm from the floor. Release the pendulum. Measure and record the distance the box moves in your Science Journal. Repeat two more times.

4. Repeat step 3 using pendulum heights of 30 cm and 45 cm.

5. Repeat steps 3 and 4 with two washers, then with three washers.

Apply It

6. How does the gravitational potential energy depend on the pendulum's weight and height?

7. How does the distance the box travels depend on the initial gravitational potential energy?

8. Does the pendulum do work on the box? Explain your answer.

9. What energy transformation occurred that caused the box to stop moving?

10. 🔑 **Key Concept** Describe the energy transformations and transfers that took place as the pendulum fell, as the pendulum hit the box, and as the box slid along the floor.

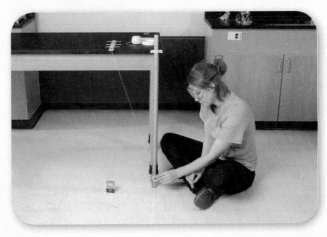

Reading Guide

Key Concepts

ESSENTIAL QUESTIONS

- What are nonrenewable energy resources?
- What are renewable energy resources?
- Why is it important to conserve energy?

Vocabulary

nonrenewable energy resource p. 178

fossil fuel p. 178

renewable energy resource p. 180

inexhaustible energy resource p. 181

g Multilingual eGlossary

Energy Resources

inquiry Extracting Energy?

Where does the electric energy come from when you turn on the lights in your home? The answer to that question depends on where you live. Different energy resources are used in different parts of the United States.

How are energy resources different?

Is there an infinite supply of usable energy, or could we someday run out of energy resources? In this activity, the red beans represent an energy resource that is available in limited amounts. The white beans represent an energy resource that is available in unlimited amounts.

1. Read and complete a lab safety form.
2. Place **40 red beans** and **40 white beans** in a **paper bag.** Mix the contents of the bag.
3. Each team should remove 20 beans from the bag without looking at the beans. Record the numbers of red and white beans in your Science Journal.
4. Put the red beans aside. They are "used up." Return all the white beans to the bag. Mix the beans in the bag. Repeat steps 3 and 4 three more times.

Think About This

1. What happened to the number of red beans drawn during each round?

2. What would eventually happen to the red beans in the bag?

3. 🔑 **Key Concept** How would changing the number of beans drawn in each round make the red beans last longer? Explain your answer.

Sources of Energy

Every day, you use many forms of energy in many ways. According to the law of conservation of energy, energy cannot be created or destroyed. Energy can only change form. Where does all the energy that you use come from?

Almost all the energy you use can be traced back to the Sun, as shown in **Figure 10.** For example, the chemical energy in the food you eat originally came from the Sun. The energy in fuels, such as gasoline, coal, and wood, also came from the Sun. In addition, a small amount of energy that reaches Earth's surface comes from inside Earth. However, the amount of energy that comes from the Sun each day is about 5,000 times greater than the amount of energy that comes from inside Earth.

Figure 10 The energy contained in food, gasoline, and wood originally came from the Sun.

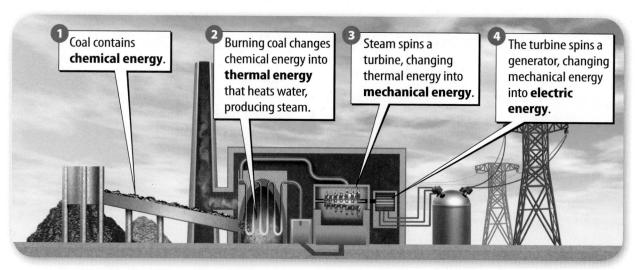

Figure 11 This coal-burning electric power plant transforms chemical energy stored in a fossil fuel into electric energy.

Visual Check What form of energy changes water into steam?

WORD ORIGIN · · · · · · · · · ·

fossil
from Latin *fossilis*, means "dug up"

Electric Power Plants

Most of the energy you use every day does not come directly from the Sun. Instead, much of the energy you use every day is in the form of electric energy. Most of the electric energy you use comes from electric power plants.

An electric power plant transforms the energy in an energy source into electric energy. There are three main energy sources used in power plants. One source of energy comes from burning fuels, such as coal. The power plant shown in **Figure 11** uses coal as an energy source. Nuclear power plants use the nuclear energy contained in uranium. Hydroelectric power plants convert the kinetic energy in falling water into electric energy.

Nonrenewable Energy Resources

The coal that a power plant burns is an example of a nonrenewable energy resource. A **nonrenewable energy resource** *is an energy resource that is available in limited amounts or that is used faster than it is replaced in nature.*

Fossil Fuels

The most commonly used nonrenewable energy resources are fossil fuels. **Fossil fuels** *are the remains of ancient organisms that can be burned as an energy source.* Fossil fuels take millions of years to form. They are being used up much faster than they form. Three types of fossil fuels are coal, natural gas, and petroleum.

Key Concept Check Why are fossil fuels considered a nonrenewable energy resource?

Formation of Petroleum 🔑

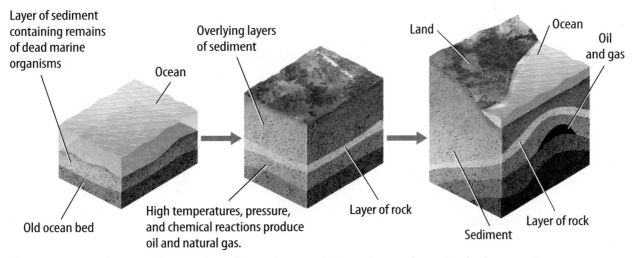

Layer of sediment containing remains of dead marine organisms

Ocean

Overlying layers of sediment

Land

Ocean

Oil and gas

Old ocean bed

High temperatures, pressure, and chemical reactions produce oil and natural gas.

Layer of rock

Sediment

Layer of rock

Figure 12 Petroleum and natural gas formed over millions of years from dead microscopic ocean organisms. Geologic processes buried these dead organisms under layers of sediment and rock. There, high temperature and pressure changed them into oil and natural gas.

The Formation of Fossil Fuels

The processes that formed fossil fuels began at Earth's surface. Petroleum and natural gas formed from microscopic ocean organisms that died and sank to the ocean floor, as shown in **Figure 12.** These organisms were gradually buried under layers of sediment—sand and mud—and rock.

Over many millions of years, increasing temperature and pressure from the weight of the sediment and rock layers changed the dead organisms into petroleum and natural gas. Coal formed on land from plants that died millions of years ago and were buried under thick layers of sediment and rock.

Using Fossil Fuels

Fossil fuels formed from organisms that changed radiant energy from the Sun to chemical potential energy. The chemical potential energy stored in fossil fuels changes to thermal energy when fossil fuels burn.

Using Petroleum Gasoline, fuel oil, diesel, and kerosene are made from petroleum. These fuels are burned mainly to power cars, trucks, and planes and to heat buildings. Petroleum is also used as a raw material in making plastics and other materials.

Using Coal Electric power plants burn about 90 percent of the coal used in the United States. Coal is also used to directly heat buildings and to produce steel and concrete. Burning coal produces more pollutants than burning other fossil fuels. In some places, these pollutants react with water vapor in the air and create acid rain. Acid rain can harm organisms, such as trees and fish.

 Reading Check What percentage of the coal used in the United States is burned in electric power plants?

Using Natural Gas About half of all homes in the United States use natural gas for heating. Electric power plants burn about 30 percent of the natural gas used in the United States. Burning natural gas produces less pollution than burning other fossil fuels.

Fossil Fuels and Global Warming

Burning fossil fuels releases carbon dioxide gas into Earth's atmosphere. Carbon dioxide is one of the gases that helps keep Earth's surface warm. However, over the past 100 years, Earth's surface has warmed by about 0.7°C. Some of this warming is due to the increasing amount of carbon dioxide produced by burning fossil fuels.

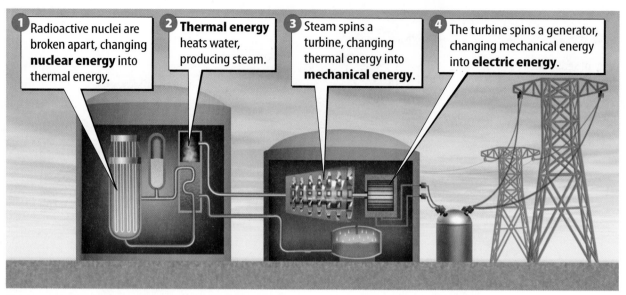

① Radioactive nuclei are broken apart, changing **nuclear energy** into thermal energy.

② **Thermal energy** heats water, producing steam.

③ Steam spins a turbine, changing thermal energy into **mechanical energy**.

④ The turbine spins a generator, changing mechanical energy into **electric energy**.

Figure 13 A nuclear power plant transforms nuclear energy into electric energy.

REVIEW VOCABULARY · · · · ·

nuclei
plural form of nucleus; the positively charged center of an atom that contains protons and neutrons.

Math Skills

Solve a One-Step Equation

Electric energy is often measured in units called *kilowatt-hours* (kWh). To calculate the electric energy used by an appliance in kWh, use this equation:

kWh = (watts/1,000) × hours

Appliances typically have a power rating measured in *watts*.

Practice

A hair dryer is rated at 1,200 watts. If you use the dryer for 0.25 hours, how much electric energy do you use?

🔲 **Review**

- **Math Practice**
- **Personal Tutor**

Nuclear Energy

Humans can also transform the nuclear energy from uranium **nuclei** into thermal energy. Uranium is found in certain minerals, but significant amounts of uranium are no longer being formed inside Earth. As a result, nuclear energy from uranium is a nonrenewable energy resource.

✓ **Reading Check** Why is nuclear energy released from uranium nuclei considered a nonrenewable energy resource?

Nuclear Power Plants In a nuclear power plant, breaking apart uranium nuclei transforms nuclear energy into thermal energy. As shown in **Figure 13,** this thermal energy changes water into the steam that spins the turbine. Unlike fossil fuel power plants, a nuclear power plant does not release pollutants into the air. However, a nuclear power plant does produce harmful nuclear waste.

Storing Nuclear Waste Nuclear waste contains radioactive materials that can harm living things. Some of these materials remain radioactive for thousands of years. Almost all nuclear waste in the United States is currently stored at the nuclear power plants where this waste is produced.

Renewable Energy Resources

Fossil fuels and uranium are being used up faster than they are being replaced. However, there are other energy resources that are not being used up. A **renewable energy resource** is *an energy resource that is replaced as fast as, or faster than, it is used.*

🔑 **Key Concept Check** Contrast renewable and nonrenewable energy resources.

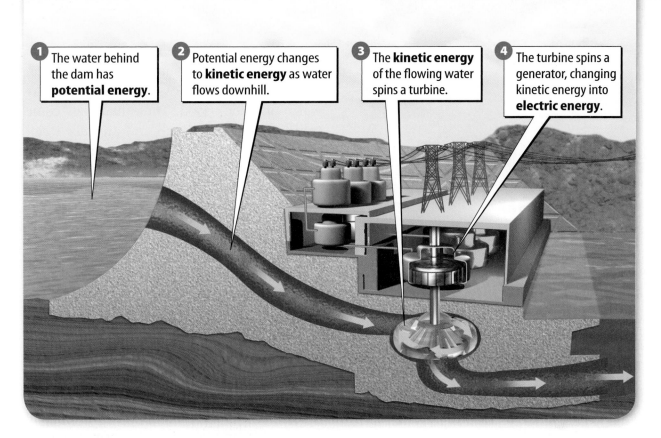

1. The water behind the dam has **potential energy**.

2. Potential energy changes to **kinetic energy** as water flows downhill.

3. The **kinetic energy** of the flowing water spins a turbine.

4. The turbine spins a generator, changing kinetic energy into **electric energy**.

Hydroelectric Power Plants

The most widely used renewable energy resource is falling water. To generate electric energy from falling water, a dam is built across a river, forming a reservoir (REH zuh vwor). As water falls through tunnels in the dam, the water's potential energy transforms into kinetic energy. **Figure 14** shows how a hydroelectric power plant transforms the kinetic energy in falling water into electric energy. Hydroelectric power plants do not emit pollutants. However, in some places, dams can disturb the life cycle of some wildlife, such as fish.

Solar Energy

Another renewable energy source is radiant energy from the Sun—solar energy. Because the Sun will produce energy for billions of years, solar energy is also an inexhaustible energy resource. An **inexhaustible energy resource** *is an energy resource that cannot be used up.* Less than about 0.1 percent of the energy used in the United States comes directly from the Sun.

Solar energy is converted directly into electric energy by solar cells. Solar cells contain materials that transform radiant energy into electric energy when sunlight strikes the solar cell. Solar cells can be placed on the roof of a building to provide electric energy, as shown in **Figure 15.**

▲ **Figure 14** A hydroelectric power plant converts the potential energy of the water stored behind the dam to electric energy.

◉ **Visual Check** Infer what would happen to the turbine if the river stopped flowing.

◉ **Concepts in Motion**
Animation

Figure 15 The dark blue panels on this roof are made of materials that convert solar energy into electric energy. ▼

▲ **Figure 16** Wind turbines convert the kinetic energy in wind into electric energy. Some wind turbines are over 60 m high.

Concepts in Motion Animation

Wind Energy

Wind energy is another inexhaustible energy resource. Modern wind turbines, such as the ones shown in **Figure 16,** convert the kinetic energy in wind into electric energy. Wind spins a propeller that is connected to an electric generator.

Wind turbines produce no pollution. However, wind turbines are practical only in regions where the average wind speed is more than about 5 m/s. Also, many wind turbines covering a large area are needed to obtain as much electric energy as one fossil fuel–burning power plant.

Biomass

People have often burned materials such as wood, dried peat moss, and manure to stay warm and cook food. These materials come from plants and animals and are called biomass. Because plant and animal materials can be replaced as fast as they are used, biomass is a renewable energy resource.

Some biomass is converted into fuels that can be burned in the engines of cars and other vehicles. Fuels made from biomass are often called biofuels. Using biofuels in vehicles can reduce the use of gasoline and make the supply of petroleum last longer.

Geothermal Energy

Thermal energy from inside Earth is called geothermal energy. This energy comes from the decay of radioactive nuclei deep inside Earth. In some places, geothermal energy produces underground pockets of hot water and steam. These pockets are called geothermal reservoirs.

In a few places, wells can be drilled to reach geothermal reservoirs. **Figure 17** shows a geothermal power plant. The hot water and steam in the geothermal reservoir are piped to the surface where they spin a turbine attached to an electric generator.

Figure 17 A geothermal power plant transforms the thermal energy from inside Earth into electric energy.

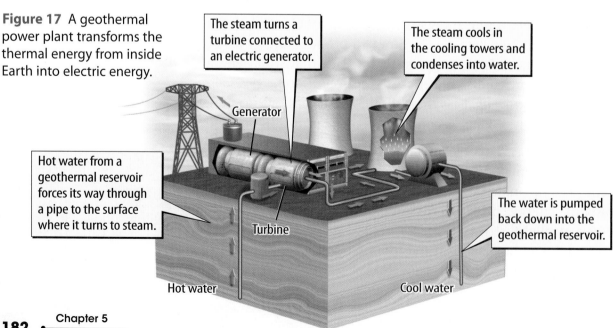

The steam turns a turbine connected to an electric generator.

The steam cools in the cooling towers and condenses into water.

Generator

Hot water from a geothermal reservoir forces its way through a pipe to the surface where it turns to steam.

The water is pumped back down into the geothermal reservoir.

Turbine

Hot water

Cool water

Sources of Energy Used in the U.S. in 2006

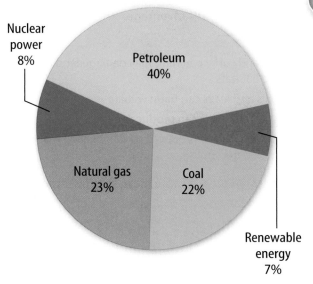

Figure 18 About 93 percent of the energy used in the United States comes from nonrenewable energy resources—fossil fuels and nuclear energy.

Conserving Energy Resources

The graph in **Figure 18** shows that fossil fuels provide about 85 percent of the energy used in the United States. Because fossil fuels are a nonrenewable energy resource, the supply decreases as they are used.

Because the supply of fossil fuels is decreasing, there could be shortages of fossil fuels in the future. Conserving energy is one way to reduce the rate at which all energy resources are used. Conserving energy means to avoid wasting energy. For example, turning off the lights when no one is in a room is a way to conserve energy.

 Key Concept Check How does conserving energy affect the rate at which energy resources are used?

In the future, energy resources besides fossil fuels might become more widely used. However, as **Table 2** on the next page shows, all energy resources have advantages and disadvantages. Comparing advantages and disadvantages will help determine which energy resources are used in the future.

Table 2 Advantages and Disadvantages of Energy Resources 🔑

Energy Resource	Advantages	Disadvantages
Nonrenewable Energy Resources		
Fossil Fuels	• Easy to transport • Widely available • Relatively inexpensive • Fossil fuel power plants are relatively inexpensive to operate.	• Drilling and surface mining may damage land and wildlife habitats. • Oil spills and leaks can harm wildlife. • Burning fossil fuels can produce air pollution. • Burning fossil fuels produces carbon dioxide that can cause global warming.
Nuclear Energy	• Nuclear power plants are relatively inexpensive to operate. • Does not produce air pollution	• Produces radioactive waste that is difficult to store • Accidents can result in dangerous leaks of radiation. • Nuclear power plants are relatively expensive to build.
Renewable Energy Resources		
Hydroelectric	• Does not pollute the air or water • Hydroelectric power plants are relatively inexpensive to operate.	• Dams damage wildlife habitats. • Dams can affect water quality and reduce the flow of water downstream. • Droughts can affect hydroelectric power plants.
Solar	• Does not pollute the air or water • Theoretically inexhaustible supply	• The amount of solar energy that reaches Earth's surface varies, depending on the location, time of day, season, and weather conditions. • A large area is needed to collect enough solar energy for a solar power plant to be viable.
Wind	• Does not pollute the air or water • Can be used in isolated areas where electricity is unavailable	• Wind turbines can be noisy. • Can be disruptive to wildlife • Only generates electricity when the wind is blowing
Geothermal	• Does not pollute the air • Geothermal power plants are relatively inexpensive to operate.	• Geothermal reservoirs are located primarily in the western United States, Alaska, and Hawaii. • Some geothermal plants produce solid wastes that require careful disposal.
Biomass	• Biofuels could replace petroleum fuels in most vehicles.	• Energy from fossil fuels is used to grow biomass. • Some farm land is used to grow biomass instead of food crops. • Burning some biomass produces pollutants, such as smoke.

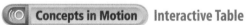

(⦿ Concepts in Motion   Interactive Table

Lesson 3 Review

Visual Summary

Nonrenewable energy resources, such as fossil fuels, are used faster than they are replaced in nature.

Renewable energy resources, such as wind energy, are replaced in nature as fast as they are used.

Conserving energy, such as driving fuel-efficient cars, is one way to reduce the rate at which energy resources are used.

FOLDABLES®

Use your lesson Foldable to review the lesson. Save your Foldable for the project at the end of the chapter.

What do you think NOW?

You first read the statements below at the beginning of the chapter.

5. Nuclear power plants release many dangerous pollutants into the air as they transform nuclear energy into electric energy.

6. Thermal energy from within Earth can be transformed into electric energy at a power plant.

Did you change your mind about whether you agree or disagree with the statements? Rewrite any false statements to make them true.

Use Vocabulary

1. **Define** *fossil fuel* in your own words.

Understand Key Concepts 🔑

2. **Compare and contrast** fossil fuels and biofuels.

3. **Explain** Conserving energy makes which type of energy resources last longer?

4. Which of these energy resources is effectively inexhaustible?
   A. biomass   C. nuclear energy
   B. fossil fuels   D. solar energy

Interpret Graphics

5. **Compare and Contrast** Copy and fill in the graphic organizer below.

Energy Resource	Similarities	Differences
Coal		
Wind		

6. **Sequence** Copy and fill in the graphic organizer below to sequence the energy transformations that occur in a coal-burning electric power plant.

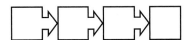

Critical Thinking

7. **Recommend** Fossil fuels form over millions of years. Use this information to explain why fossil fules are a nonrenewable energy resource.

Math Skills ×÷+

📖 Review —Math Practice—

8. A 22-W compact fluorescent lightbulb (CFL) produces as much light as a 100-W regular lightbulb. How much electric energy in kWh does each bulb use in 10 hours?

Pinwheel Power

Materials

round pencil with unused eraser

metal washers

cardboard container

sand or small rocks

three-speed hair dryer

stopwatch

Also needed: manila folder, metric ruler, scissors, single-hole punch, thread

Safety

Moving air, or wind, is a renewable energy resource. In some places, wind turbines transform the kinetic energy of wind into electric energy. This electric energy can be used to do work by making an object move. In this lab, you will construct a pinwheel turbine and observe how changes in wind speed affect the rate at which your wind turbine does work.

Ask a Question

How does the wind speed affect the rate at which a wind turbine does work?

Make Observations

1. Read and complete a lab safety form.
2. Construct a pinwheel from a manila folder using the diagram below as a guide.
3. Use the plastic push pin to carefully attach the pinwheel to the eraser of the pencil.
4. Use the hole punch to make holes on opposite sides of the top of the container. Use your ruler to make sure the holes are exactly opposite one another. Weigh down the container with sand or small rocks.
5. Put the pencil through the holes and make sure it spins freely. Blow against the blades of the pinwheel with varying amounts of force to observe how the pinwheel moves. Record your observations in your Science Journal.
6. Measure and cut 100 cm of thread. Tie the washers to one end of the thread. Tape the other end of the thread to the pencil. Your wind turbine should resemble the one shown on the next page.

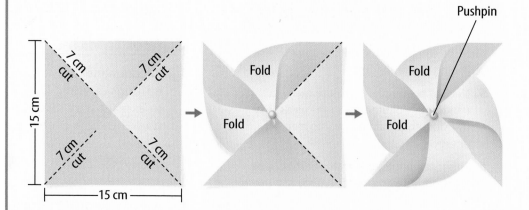

Form a Hypothesis

7 Use your observations from step 5 to form a hypothesis about the relationship between wind speed and the rate at which the wind turbine does work.

Test Your Hypothesis

8 Work with two other students to test your hypothesis. One person will use the hair dryer to model a "slow" wind speed. Another person will stop the pencil's movement after 5 seconds on the stopwatch. The third person will measure the length of thread remaining between the pencil and the top of the washers. Then, someone will unwind the thread and the group will repeat this procedure four more times with the dryer on low. Record all data in your Science Journal.

9 Repeat step 8 with the dryer on medium.

10 Repeat step 8 with the dryer on high.

Analyze and Conclude

11 **Interpret Data** Did your hypothesis agree with your data and observations? Explain.

12 **Sequence** Describe how energy was transformed from one form into another in this lab.

13 **Draw Conclusions** What factors might have affected the rate at which your pinwheel turbine did work?

14 **The Big Idea** Explain how wind is used as an energy resource.

Communicate Your Results

Use your data and observations to write a paragraph explaining how wind speed affects the rate at which a wind turbine can do work.

8

Inquiry Extension

Research designs of real wind generators and then create a model of one. Write a short explanation of its advantages and disadvantages compared to other real wind generators.

Lab Tips

☑ You measure the rate at which the wind turbine does work by measuring how fast the turbine lifts the metal washers.

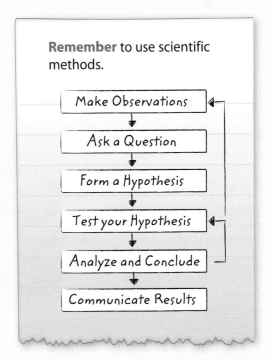

Remember to use scientific methods.

Make Observations

Ask a Question

Form a Hypothesis

Test your Hypothesis

Analyze and Conclude

Communicate Results

Chapter 5 Study Guide

 THE BIG IDEA Energy is the ability to cause change. Energy resources contain energy that can be transformed into other, more useful forms of energy.

Key Concepts Summary 🔑

Vocabulary

Lesson 1: Forms of Energy

- **Energy** is the ability to cause change.
- **Kinetic energy** is the energy a body has because it is moving. **Potential energy** is stored energy.
- Different forms of energy include **thermal energy** and **radiant energy.**

energy p. 161
kinetic energy p. 162
potential energy p. 162
work p. 164
mechanical energy p. 165
sound energy p. 165
thermal energy p. 165
electric energy p. 165
radiant energy p. 165
nuclear energy p. 165

Lesson 2: Energy Transformations

- Any form of energy can be transformed into other forms of energy.
- According to the **law of conservation of energy,** energy can be transformed from one form into another or transferred from one region to another, but energy cannot be created or destroyed.
- **Friction** transforms mechanical energy into thermal energy.

law of conservation of energy p. 170
friction p. 171

Lesson 3: Energy Resources

- A **nonrenewable energy resource** is an energy resource that is available in a limited amount and can be used up.
- A **renewable energy resource** is replaced in nature as fast as, or faster, than it is used.
- Conserving energy, such as turning off lights when they are not needed, is one way to reduce the rate at which energy resources are used.

nonrenewable energy resource p. 178
fossil fuel p. 178
renewable energy resource p. 180
inexhaustible energy resource p. 181

FOLDABLES® Chapter Project

Assemble your Lesson Foldables as shown to make a Chapter Project. Use the project to review what you have learned in this chapter.

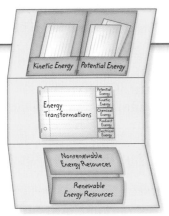

Use Vocabulary

Each of the following sentences is false. Make the sentence true by replacing the italicized word with a vocabulary term.

1 *Thermal energy* is the form of energy carried by an electric current.

2 The *chemical potential energy* of an object depends on its mass and speed.

3 *Friction* is the transfer of energy that occurs when a force is applied over a distance.

4 Natural gas is considered *an inexhaustible energy resource.*

5 *Radiant energy* is energy that is stored in the nucleus of an atom.

Link Vocabulary and Key Concepts

🔘 **Concepts in Motion**   Interactive Concept Map

Copy this concept map, and then use vocabulary terms from the previous page to complete the concept map.

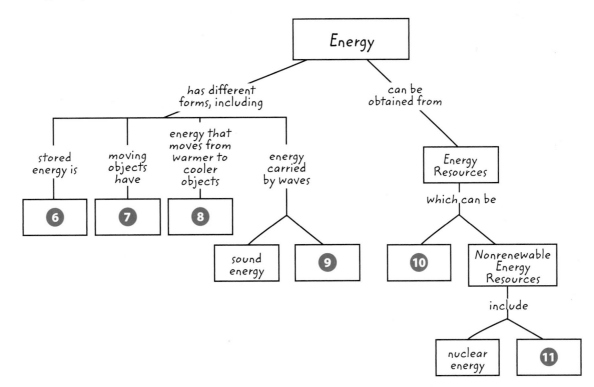

Understand Key Concepts

1. What factors determine an object's kinetic energy?
   A. its height and mass
   B. its mass and speed
   C. its size and weight
   D. its speed and height

2. The gravitational potential energy stored between an object and Earth depends on
   A. the object's height and weight.
   B. the object's mass and speed.
   C. the object's size and weight.
   D. the object's speed and height.

3. When a ball is thrown upward, where does it have the least kinetic energy?
   A. at its highest point
   B. at its lowest point when it is moving downward
   C. at its lowest point when it is moving upward
   D. midway between its highest point and its lowest point

4. What energy transformation occurs in the panels on this roof?

   A. chemical energy to thermal energy
   B. nuclear energy to electric energy
   C. radiant energy to electric energy
   D. sound energy to thermal energy

5. According to the law of conservation of energy, which is always true?
   A. Almost all energy used on Earth comes from fossil fuels.
   B. Energy can never be created or destroyed.
   C. Nuclear energy is a renewable energy resource.
   D. Work is done when a force is exerted on an object.

6. Which energy resource is being formed in the picture below?

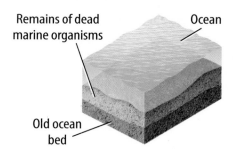

   A. biomass
   B. geothermal reservoirs
   C. petroleum
   D. uranium

7. In which situation would the gravitational potential energy between you and Earth be greatest?
   A. You are running down a hill.
   B. You are running up a hill.
   C. You stand at the bottom of a hill.
   D. You stand at the top of a hill.

8. Which produces the most air pollutants?
   A. burning coal
   B. burning natural gas
   C. hydroelectric power plants
   D. nuclear power plants

9. Which is an example of a renewable energy resource?
   A. geothermal energy
   B. natural gas
   C. nuclear energy
   D. petroleum

10. Which best describes coal?
    A. It burns without polluting the air.
    B. It formed from the remains of plants.
    C. It is a renewable energy resource.
    D. It is the energy source used at nuclear power plants.

Critical Thinking

11 **Determine** if work is done on the nail if a person pulls the handle to the left and the hammer moves. Explain your reasoning.

12 **Contrast** the energy transformations that occur in an electric toaster oven and in an electric fan.

13 **Infer** A red box and a blue box are on the same shelf. There is more gravitational potential energy between the red box and Earth than between the blue box and Earth. Which box weighs more? Explain your answer.

14 **Infer** Juanita moves a round box and a square box from a lower shelf to a higher shelf. The gravitational potential energy for the round box increases by 50 J. The gravitational potential energy for the square box increases by 100 J. On which box did Juanita do more work? Explain your reasoning.

15 **Explain** why a skateboard coasting on a flat surface slows down and comes to a stop.

16 **Describe** the difference between the law of conservation of energy and what is meant by conserving energy.

17 **Decide** Harold stretches a rubber band and lets it go. The rubber band flies across the room. He says this demonstrates the transformation of kinetic energy to elastic potential energy. Is Harold correct? Explain.

Writing in Science

18 **Write** a short essay explaining which energy resources you think will be most important in the future.

REVIEW THE BIG IDEA

19 Write an explanation of energy and energy resources for a fourth grader who has never heard of these terms.

20 Identify five energy transformations in the photo below.

Math Skills

Review

Math Practice

21 An electric water heater is rated at 5,500 watts and operates for 106 hours per month. How much electric energy in kWh does the water heater use each month?

22 A family uses 1,303 kWh of electric energy in a month. If the power company charges $0.08 cents per kilowatt hour, what is the total electric energy bill for the month?

Record your answers on the answer sheet provided by your teacher or on a sheet of paper.

Multiple Choice

1 Which is the transfer of energy that occurs when a player throws a basketball toward a hoop?

A displacement

B force

C velocity

D work

Use the diagram below to answer questions 2 and 3.

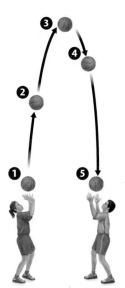

2 At which points is the kinetic energy of the basketball greatest?

A 1 and 5

B 2 and 3

C 2 and 4

D 3 and 4

3 At which point is the gravitational potential energy at a maximum?

A 1

B 2

C 3

D 4

Use the table below to answer question 4.

Vehicle	Mass	Speed
Car 1	1,200 kg	20 m/s
Car 2	1,500 kg	20 m/s
Truck 1	4,800 kg	20 m/s
Truck 2	6,000 kg	20 m/s

4 Which vehicle has the most kinetic energy?

A car 1

B car 2

C truck 1

D truck 2

5 Which form of energy is most abundant on Earth?

A geothermal

B fossil fuel

C solar

D wind

6 Which energy source has been linked to global warming?

A coal

B geothermal

C hydroelectric

D wind

7 A bicyclist uses brakes to slow from 3 m/s to a stop. What stops the bike?

A cohesion

B acceleration

C friction

D gravity

Use the diagram below to answer question 8.

8 The work being done in the diagram above transfers energy to

A the box.

B the floor.

C the girl.

D the shelf.

9 Which is true of energy?

A It cannot be created or destroyed.

B It cannot change form.

C Most forms cannot be conserved.

D Most forms cannot be traced to a source.

10 From which does biomass originate?

A animal and plant material

B nuclear fission

C manufacturing or industrial waste

D power plants

Constructed Response

Use the table below to answer questions 11 and 12.

Forms of Energy	Definition

11 List six forms of energy in the table. Briefly define each form.

12 Provide real-life examples of each of the listed forms of energy.

Use the diagram below to answer question 13.

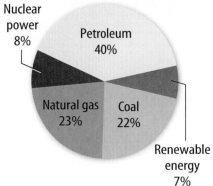

Sources of Energy Used in the U.S. in 2006

Nuclear power 8%

Petroleum 40%

Natural gas 23%

Coal 22%

Renewable energy 7%

13 What percentage of U.S. energy usage comes from renewable resources? Why is energy conservation important?

NEED EXTRA HELP?													
If You Missed Question...	1	2	3	4	5	6	7	8	9	10	11	12	13
Go to Lesson...	1	2	2	1	3	3	2	1	2	3	1	2	3

Thermal Energy

THE BIG IDEA

How can thermal energy be used?

Inquiry What are these colors?

This image shows the thermal energy of cars moving in traffic. The white indicates regions of high thermal energy, and the dark blue indicates regions of low thermal energy.

- What is thermal energy?
- How does thermal energy relate to temperature and heat?
- How can thermal energy be used?

Get Ready to Read

What do you think?

Before you read, decide if you agree or disagree with each of these statements. As you read this chapter, see if you change your mind about any of the statements.

1. Temperature is the same as thermal energy.

2. Heat is the movement of thermal energy from a hotter object to a cooler object.

3. It takes a large amount of energy to significantly change the temperature of an object with a low specific heat.

4. The thermal energy of an object can never be increased or decreased.

5. Car engines create energy.

6. Refrigerators cool food by moving thermal energy from inside the refrigerator to the outside.

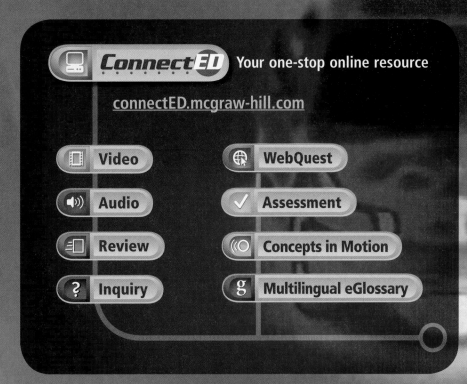

ConnectED Your one-stop online resource

connectED.mcgraw-hill.com

- Video
- Audio
- Review
- Inquiry
- WebQuest
- Assessment
- Concepts in Motion
- Multilingual eGlossary

Thermal Energy, Temperature, and Heat

Reading Guide

Key Concepts 🔑
ESSENTIAL QUESTIONS

- How are temperature and kinetic energy related?
- How do heat and thermal energy differ?

Vocabulary

thermal energy p. 198

temperature p. 199

heat p. 201

g   **Multilingual eGlossary**

Inquiry   How hot is it?

Forty gallons of sugar-maple sap must be heated to a very high temperature for several days to produce 1 gallon of maple syrup. What kind of energy is needed to achieve this very high temperature? Is there a difference between heat, temperature, and thermal energy?

How can you describe temperature?

Have you ever used Fahrenheit or Celsius to describe the temperature? Why can't you just make up your own temperature scale?

1. Read and complete a lab safety form.

2. Use a **ruler** and a **permanent marker** to divide a **clear plastic straw** into 12 equal parts. Number the lines. Give your scale a name.

3. Add a room-temperature **colored alcohol-water mixture** to an **empty plastic water bottle** until it is about $\frac{1}{4}$ full.

4. Place one end of the straw into the bottle with the tip just below the surface of the liquid. Seal the straw onto the bottle top with **clay.**

5. Place the bottle in a **hot water bath**, and observe the liquid in your straw.

Think About This

1. Why is it important for scientists to use the same scale to measure temperature?

2. **Key Concept** What are some ways to make the liquid in your thermometer rise or fall?

Kinetic and Potential Energy

What do a soaring soccer ball and the particles that make up hot maple syrup have in common? They all have energy, or the ability to cause change. What type of energy does a moving soccer ball have? Recall that any moving object has kinetic energy. When the athlete in **Figure 1** kicks the ball and puts it in motion, the ball has kinetic energy.

In addition to kinetic energy, when the soccer ball is in the air, it has potential energy. Potential energy is stored energy due to the interaction between two objects. For example, think of Earth as one object and the ball as another. When the ball is in the air, it is attracted to Earth due to gravity. This attraction is called gravitational potential energy. In other words, since the ball has the potential to change, it has potential energy. And, the higher the ball is in the air, the greater the potential energy of the ball.

You also might recall that the potential energy plus the kinetic energy of an object is the mechanical energy of the object. When a soccer ball is flying through the air, you could describe the mechanical energy of the ball by describing both its kinetic and potential energy. On the next page, you will read about how the particles that make up maple syrup have energy, just like a soaring soccer ball.

Reading Check How could you describe the energy of a moving object?

Figure 1 This soccer ball has both kinetic energy and potential energy.

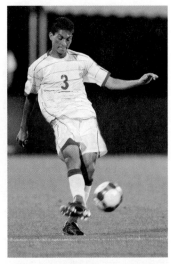

What is thermal energy?

Every solid, liquid, and gas is made up of trillions of tiny particles that are constantly moving. Moving particles make up the books you read, the air you breathe, and the maple syrup you pour on your pancakes. For example, the particles that make up a book, or any solid, vibrate in place. The particles that make up the air around you, or any gas, are spread out and move freely and quickly. Because particles are in motion, they have kinetic energy, like the soaring soccer ball in **Figure 2.** The faster particles move, the more kinetic energy they have.

The particles that make up matter also have potential energy. Like the interaction between a soccer ball and Earth, particles that make up matter interact with and are attracted to one another. The particles that make up solids usually are held very close together by attractive forces. The particles that make up a liquid are slightly farther apart than those that make up a solid. And, the particles that make up a gas are much more spread out than those that make up either a solid or a liquid. The greater the average distance between particles, the greater the potential energy of the particles.

Recall that a flying soccer ball has mechanical energy, which is the sum of its potential energy and its kinetic energy. The particles that make up the soccer ball, or any material, have a similar kind of energy called thermal energy. **Thermal energy** *is the sum of the kinetic energy and the potential energy of the particles that make up a material.* Thermal energy describes the energy of the particles that make up a solid, a liquid, or a gas.

Reading Check How are thermal energy and mechanical energy similar? How are they different?

Figure 2 The potential energy of the soccer ball depends on the distance between the ball and Earth. The potential energy of the particles of matter depends on their distance from one another.

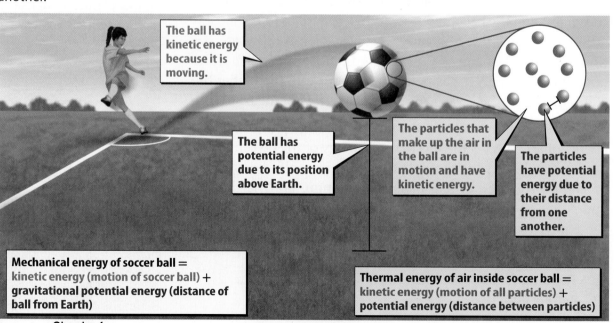

The ball has kinetic energy because it is moving.

The ball has potential energy due to its position above Earth.

The particles that make up the air in the ball are in motion and have kinetic energy.

The particles have potential energy due to their distance from one another.

Mechanical energy of soccer ball = kinetic energy (motion of soccer ball) + gravitational potential energy (distance of ball from Earth)

Thermal energy of air inside soccer ball = kinetic energy (motion of all particles) + potential energy (distance between particles)

Figure 3 The air's temperature depends on how fast the particles in the air move.

✓**Visual Check** What happens to the motion of the particles in the air as temperature increases?

Concepts in Motion
Animation

What is temperature?

When you think of temperature, you probably think of it as a measurement of how warm or cold something is. However, scientists define temperature in terms of kinetic energy.

Average Kinetic Energy and Temperature

The particles that make up the air inside and outside the house in **Figure 3** are moving. However, they are not moving at the same speed. The particles in the air in the warm house move faster and have more kinetic energy than those outside on a cold winter evening. **Temperature** *represents the average kinetic energy of the particles that make up a material.*

The greater the average kinetic energy of particles, the greater the temperature. The temperature of the air inside the house is higher than the temperature of the air outside the house. This is because the particles that make up the air inside the house have greater average kinetic energy than those outside. In other words, the particles of air inside the house are moving at a greater average speed than those outside.

Key Concept Check How are temperature and kinetic energy related?

Thermal Energy and Temperature

Temperature and thermal energy are related, but they are not the same. For example, as a frozen pond melts, both ice and water are present and they have the same temperature. Therefore, the particles that make up the ice and the water have the same average kinetic energy, or speed. However, the particles do not have the same thermal energy. This is because the average distance of the particles that make up liquid water and ice are different. The particles that make up the liquid and the solid water have different potential energies and, therefore, different thermal energies.

WORD ORIGIN

temperature
from Latin *temperatura*, means "moderating, tempering"

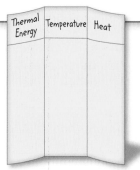

FOLDABLES

Make a vertical three-column chart book. Label it as shown. Use it to organize your notes on the properties of heat, temperature, and thermal energy.

Thermal Energy	Temperature	Heat

Measuring Temperature

How can you measure temperature? It would be impossible to measure the kinetic energy of individual particles and then calculate their average kinetic energy to determine the temperature. Instead, you can use thermometers, such as the ones in **Figure 4,** to measure temperature.

A common type of thermometer is a bulb thermometer. A bulb thermometer is a glass tube connected to a bulb that contains a liquid such as alcohol. When the temperature of the alcohol increases, the alcohol expands and rises in the glass tube. When the temperature of the alcohol decreases, the alcohol contracts back into the bulb. The height of the alcohol in the tube indicates the temperature.

There are other types of thermometers too, such as an electronic thermometer. This thermometer measures changes in the resistance of an electric circuit and converts this measurement to a temperature.

Temperature Scales

You might have seen the temperature in a weather report given in degrees Fahrenheit and degrees Celsius. On the Fahrenheit scale, water freezes at 32° and boils at 212°. On the Celsius scale, water freezes at 0° and boils at 100°. The Celsius scale is used by scientists worldwide.

Scientists also use the Kelvin scale. On the Kelvin scale, water freezes at 273 K and boils at 373 K. The lowest possible temperature for any material is 0 K. This is known as absolute zero. If a material were at 0 K, the particles in that material would not be moving and would no longer have kinetic energy. Scientists have not been able to cool any material to 0 K.

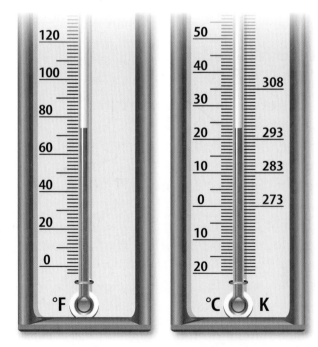

Figure 4 Thermometers measure temperature. Common temperature scales are Celsius, Kelvin, and Fahrenheit.

The hot cocoa has a high temperature. Thermal energy is transferred from the mug to its surroundings.

The heat from the hot cocoa transferred to the air is greater than the heat transferred from the hot cocoa to the girl's hands. This is because the temperature difference is greater from the hot cocoa to the air.

Figure 5 The hot cocoa heats the air and the girl's hands.

What is heat?

Have you ever held a cup of hot cocoa on a cold day like the girl in **Figure 5?** When you do, thermal energy moves from the warm cup to your hands. *The movement of thermal energy from a warmer object to a cooler object is called* **heat.** Another way to say this is that thermal energy from the cup heats your hands, or the cup is heating your hands.

Just as temperature and thermal energy are not the same thing, neither are heat and thermal energy. All objects have thermal energy. However, you heat something when thermal energy transfers from one object to another. The girl in **Figure 5** heats her hands because thermal energy transfers from the hot cocoa to her hands.

Key Concept Check How do heat and thermal energy differ?

The rate at which heating occurs depends on the difference in temperatures between the two objects. The difference in temperatures between the hot cocoa and the air is greater than the difference in temperatures between the hot cocoa and the cup. The hot cocoa heats the air more than it heats the cup. Heating continues until all objects in contact are the same temperature.

Inquiry MiniLab
10 minutes

How do temperature scales compare?

If someone told you it was 2°C or 300 K outside, would you know whether it was warm or cold?

	Celsius (°C)	Fahrenheit (°F)	Kelvin (K)
Room temperature			
Light jacket weather			
Hot summer day			

1. Copy the table into your Science Journal.
2. Lay a **ruler** across **Figure 4** so that it lines up with the temperatures at which water freezes. Record the temperatures
3. Repeat step 2 for the three values in the table.

Analyze and Conclude

1. **Estimate** Imagine that it is snowing outside. What might the temperature be in degrees Celsius? In kelvin?

2. **Key Concept** Why doesn't the Kelvin scale include negative numbers?

Lesson 1 Review

Visual Summary

The greater the distance between two particles or two objects, the greater the potential energy.

Heat is the movement of thermal energy from a warmer object to a cooler object.

When thermal energy moves between a material and its environment, the material's temperature changes.

 FOLDABLES®

Use your lesson Foldable to review the lesson. Save your Foldable for the project at the end of the chapter.

What do you think NOW?

You first read the statements below at the beginning of the chapter.

1. Temperature is the same as thermal energy.

2. Heat is the movement of thermal energy from a hotter object to a cooler object.

Did you change your mind about whether you agree or disagree with the statements? Rewrite any false statements to make them true.

Use Vocabulary

1. The sum of kinetic energy and potential energy of the particles in a material is _____.

2. **Relate** temperature to the average kinetic energy in a material.

Understand Key Concepts

3. **Differentiate** between thermal energy and heat.

4. Which increases the kinetic energy of the particles that make up a bowl of soup?
   - A. dividing the soup in half
   - B. putting the soup in a refrigerator
   - C. heating the soup for 1 min on a stove
   - D. decreasing the distance between the particles that make up the soup

5. **Infer** Suppose a friend tells you he has a temperature of 38°C. Your temperature 37°C. Do the particles that make up your body or your friend's body have a greater average kinetic energy? Explain.

Interpret Graphics

6. **Identify** Copy and fill in the following graphic organizer to show the forms of energy that make up thermal energy.

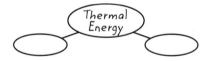

Critical Thinking

7. **Explain** How could you increase the kinetic thermal energy of a liquid?

Math Skills

Review
——— Math Practice ———

8. Maple sap boils at 104°C. At what Fahrenheit temperature does the sap boil?

How do different materials affect thermal energy transfer?

Materials

cardboard

100-mL graduated cylinder

———————

2 thermometers

Also Needed

1-L square plastic container, test containers (metal, polystyrene, ceramic, glass, plastic), large rubber band, hot water

Safety

You might have noticed that thermal energy moves more easily through some substances than others. For example, juice stays colder in a foam cup than in a can. How does the container's material affect how quickly thermal energy moves through it?

Learn It

To **form a hypothesis** is to propose an explanation for an observation. The explanation should be testable. One way to **test a hypothesis** is by gathering data that shows whether the hypothesis is correct.

Try It

1. Read and complete a lab safety form.

2. Observe the test containers. Write a hypothesis in your Science Journal that explains why you think a certain material will slow the transfer of thermal energy more than others.

3. Copy the table below.

4. Each lab group will test one container. Stand your test container in the center of a 1-L plastic container.

5. Add 125 mL of hot water to the test container. Measure and record the water's temperature.

6. Add room-temperature water to the plastic container until the level in both containers is equal. Measure and record the room-temperature water's temperature.

7. Place a cardboard square over the test container. Use two thermometers to take the temperature of the water in both containers every 2 min for 20 min. Record your data in your table.

8. Compare your data with the data gathered by the other teams. Rank the test containers from slowest to fastest thermal energy transfer in your Science Journal.

Apply It

9. **Analyze Data** Did your data support your hypothesis? Why or why not?

10. 🔑 **Key Concept** What happened to the thermal energy of the water in the test container? Why did this happen?

°C	0 min	2 min	4 min	6 min	8 min	10 min	12 min	14 min	16 min	18 min	20 min
Temp in test container											
Temp in outer container											

Lesson 2

Reading Guide

Key Concepts
ESSENTIAL QUESTIONS

- What is the effect of having a small specific heat?
- What happens to a material when it is heated?
- In what ways can thermal energy be transferred?

Vocabulary

radiation p. 205

conduction p. 206

thermal conductor p. 206

thermal insulator p. 206

specific heat p. 207

thermal contraction p. 208

thermal expansion p. 208

convection p. 210

convection current p. 211

 Multilingual eGlossary

Video Science Video

Thermal Energy Transfers

Inquiry Keeping Warm?

Imagine camping in the mountains on a cold winter night. Your survival could depend on keeping warm. There are many things you could do to get warm and stay warm. In this picture, how is thermal energy transferred from the fire to the camper? Why does his coat keep him from losing thermal energy?

Launch Lab

15 minutes

How hot is it?

When you touch an ice cube, you sense that it is cold. When you get inside a car on a warm day, you sense that it is hot. How accurate is your sense of touch in predicting temperature?

1. Read and complete a lab safety form.

2. Place the palm of one hand flat against a piece of **metal** and the other hand against a piece of **wood.** Observe which material feels colder, and record it in your Science Journal.

3. Repeat step 2 with other materials, including **cardboard, glass, plastic,** and **foam.**

4. Rank the materials from coldest to warmest in your Science Journal.

5. Place a **liquid crystal thermometer** on each material. Record the temperature of each material in your Science Journal.

Think About This

1. Were you able to accurately rank the materials by temperature only by touching them?

2. **Key Concept** Why might some of the materials in this experiment feel cooler than others even though they are in the same room?

How is thermal energy transferred?

Have you ever gotten into a car, such as the one in **Figure 6,** on a hot summer day? You can guess that the inside of the car is hot even before you touch the door handle. You open the door and hot air seems to pour out of the car. When you touch the metal safety-belt buckle, it is hot. How is thermal energy transferred between objects? Thermal energy is transferred in three ways—by radiation, conduction, and convection.

Radiation

The transfer of thermal energy from one material to another by electromagnetic waves is called **radiation.** All matter, including the Sun, fire, you, and even ice, transfers thermal energy by radiation. Warm objects emit more radiation than cold objects do. For example, when you place your hands near a fire, you can more easily feel the transfer of thermal energy by radiation than when you place your hands near a block of ice.

Thermal energy from the Sun heats the inside of the car in **Figure 6** by radiation. In fact, radiation is the only way thermal energy can travel from the Sun to Earth. This is because space is a **vacuum.** However, radiation also transfers thermal energy through solids, liquids, and gases.

✓ **Reading Check** How does the Sun heat the inside of a car?

SCIENCE USE V. COMMON USE

vacuum
Science Use a space that contains little or no matter

Common Use a device for cleaning carpets and rugs that uses suction

Figure 6 The Sun heats this car by radiation.

Lesson 2

205

EXPLORE

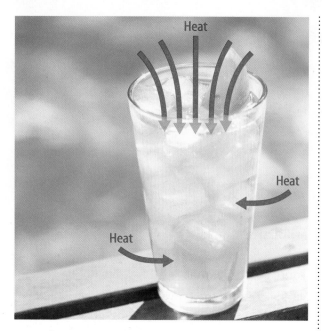

Figure 7 🗝 The hot air transfers thermal energy to, or heats, the cool lemonade by conduction. Eventually the kinetic thermal energy and temperature of the air and the lemonade will be equal.

((◉)) **Concepts in Motion** **Animation**

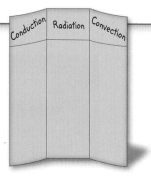
Conduction

Suppose it's a hot day and you have a cold glass of lemonade, such as the one in **Figure 7.** The lemonade has a lower temperature than the surrounding air. Therefore, the particles that make up the lemonade have less kinetic energy than the particles that make up the air. When particles with different kinetic energies collide, the particles with higher kinetic energy transfer energy to particles with lower kinetic energy.

In **Figure 7,** the particles that make up the air collide with and transfer kinetic energy to the particles that make up the lemonade. As a result, the average kinetic energy, or temperature, of the particles that make up the lemonade increases. Since kinetic energy is being transferred, thermal energy is being transferred. *The transfer of thermal energy between materials by the collisions of particles is called* **conduction.** Conduction continues until the thermal energy of all particles in contact is equal.

Thermal Conductors and Insulators

On a hot day, why does a metal safety-belt buckle in a car feel hotter than the safety belt? Both the buckle and safety belt receive the same amount of thermal energy from the Sun. The metal that makes up the buckle is a good thermal conductor. *A* **thermal conductor** *is a material through which thermal energy flows easily.* Atoms in good thermal conductors have electrons that move easily. These electrons transfer kinetic energy when they collide with other electrons and atoms. Metals are better thermal conductors than nonmetals. The cloth that makes up a safety belt is a good thermal insulator. *A* **thermal insulator** *is a material through which thermal energy does not flow easily.* The electrons in the atoms of a good thermal insulator do not move easily. These materials do not transfer thermal energy easily because fewer collisions occur between electrons and atoms.

Specific Heat

The amount of thermal energy required to increase the temperature of 1 kg of a material by 1°C is called **specific heat.** Every material has a specific heat. It does not take much energy to change the temperature of a material with a low specific heat but it can take a lot of energy to change the temperature of a material with high specific heat.

Thermal conductors, such as the cloth of the safety-belt buckle in **Figure 8,** have a lower specific heat than thermal insulators, such as the cloth safety belt. This means it takes less thermal energy to increase the buckle's temperature than it takes to increase the temperature of the cloth safety belt by the same amount.

The specific heat of water is particularly high. It takes a large amount of energy to increase or decrease the temperature of water. The high specific heat of water has many beneficial effects. For example, much of your body is water. Water's high specific heat helps prevent your body from overheating. The high specific heat of water is one of the reasons why pools, lakes, and oceans stay cool in summer. Water's high specific heat also makes it ideal for cooling machinery, such as car engines and rock cutting saws.

 Key Concept Check What does it mean if a material has a low specific heat?

Specific Heat 🔑

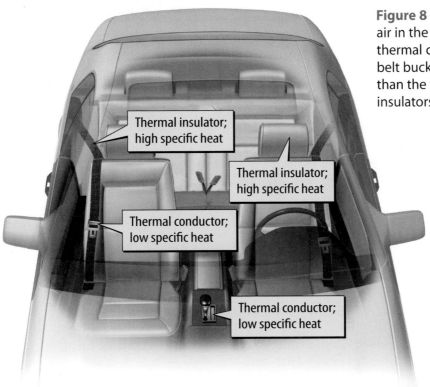

Thermal insulator; high specific heat

Thermal insulator; high specific heat

Thermal conductor; low specific heat

Thermal conductor; low specific heat

Figure 8 On a hot summer day, the air in the car is hot. The temperature of thermal conductors, such as the safety-belt buckles, increases more quickly than the temperature of thermal insulators, such as the seat material.

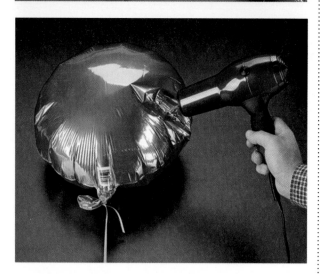

▲ **Figure 9** 🔑 Air inside the balloon increases in volume when the temperature increases.

▲ **Figure 10** 🔑 Sidewalks can withstand thermal expansion and contraction because of control joints.

Thermal Expansion and Contraction

What happens if you take an inflated balloon outside on a cold day? Thermal energy transfers from the particles that make up the air inside the balloon to the particles that make up the balloon material and then to the cold outside air. As the particles that make up the air in the balloon lose thermal energy, which includes kinetic energy, they slow down and get closer together. This causes the volume of the balloon to decrease. **Thermal contraction** *is a decrease in a material's volume when its temperature decreases.*

How could you reinflate the balloon? You could heat the air inside the balloon with a hair dryer, like in **Figure 9.** The particles that make up the hot air coming out of the hair dryer transfer thermal energy, which includes kinetic energy, to the particles that make up the air inside the balloon. As the average kinetic energy of the particles increases, the air temperature increases. Also, as the average kinetic energy of the particles increases, they speed up and spread out, increasing the volume of air inside the balloon. **Thermal expansion** *is an increase in a material's volume when its temperature increases.*

Thermal expansion and contraction are most noticeable in gases, less noticeable in liquids, and the least noticeable in solids.

 Key Concept Check What happens to the volume of a gas when it is heated?

Sidewalk Gaps

In many places, outdoor temperatures become very hot in the summer. High temperatures can cause thermal expansion in structures, such as concrete sidewalks. If the concrete expands too much or expands unevenly, it could crack. Therefore, control joints are cut into sidewalks, as shown in **Figure 10.** If the sidewalk does crack, it should crack smoothly at the control joint.

Hot-Air Balloons

How do hot-air balloons work? As shown in **Figure 11**, a burner heats the air in the balloon, causing thermal expansion. The particles that make up the air inside the balloon move faster and faster. As the particles collide with one another, some are forced outside the balloon through the opening at the bottom. Now, there are fewer particles in the balloon than in the same volume of air outside the balloon. The balloon is less dense, and it begins to rise through denser outside air.

To land a hot-air balloon, the balloonist allows the air inside the balloon to gradually cool. The air undergoes thermal contraction. However, the balloon itself does not contract. Instead, denser air from outside the balloon fills the space inside. As the density of the balloon increases, it slowly descends.

Ovenproof Glass

If you put an ordinary drinking glass into a hot oven, the glass might break or shatter. However, an ovenproof glass dish would not be damaged in a hot oven. Why is this so?

Different parts of ordinary glass expand at different rates when heated. This causes it to crack or shatter. Ovenproof glass is designed to expand less than ordinary glass when heated, which means that it usually does not crack in the oven.

Figure 11 Hot-air balloonists control their balloons using thermal expansion and contraction.

Inquiry **MiniLab**

20 minutes

How does adding thermal energy affect a wire?

How could thermal energy help you remove a metal lid from a glass jar?

1 Read and complete a lab safety form.

2 Set up **two ring stands** so that the rings are 1–2 m apart. Tie the ends of a **2-m length of wire** to the rings so that the wire is straight and tight.

3 Use **thread** to tie a **weight** to the middle of the wire.

4 Use a **ruler** to measure the distance from the bottom of the weight to the table. Record your data in your Science Journal.

5 Using **matches** light two **candles.** Move the candle flames back and forth under the wire. Repeat step 4 every minute for 5 min. Blow out the candles.

6 Repeat step 4 again every minute for 5 min as the wire cools.

Analyze and Conclude

1. **Predict** What would happen if you continued to heat the wire? Explain.

2. **Apply** How could you use this idea to help you remove a metal lid from a glass jar?

3. **Key Concept** What happens to the particles that make up the wire when the wire is heated? How do you know?

Figure 12 This cycle of cooler water sinking and forcing warmer water upward is an example of convection.

Review

Personal Tutor

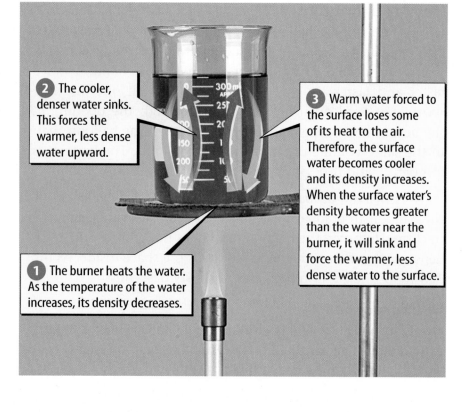

2 The cooler, denser water sinks. This forces the warmer, less dense water upward.

3 Warm water forced to the surface loses some of its heat to the air. Therefore, the surface water becomes cooler and its density increases. When the surface water's density becomes greater than the water near the burner, it will sink and force the warmer, less dense water to the surface.

1 The burner heats the water. As the temperature of the water increases, its density decreases.

Convection

When you heat a pan of water on the stove, the burner heats the pan by conduction. This process, shown in **Figure 12,** involves the movement of thermal energy within a fluid. The particles that make up liquids and gases move around easily. As they move, they transfer thermal energy from one location to another. **Convection** *is the transfer of thermal energy by the movement of particles from one part of a material to another.* Convection only occurs in fluids, such as water, air, magma, and maple syrup.

WORD ORIGIN · · · · · · · · · · · ·

convection
from Latin *convectionem,*
means "the act of carrying"
· · · · · · · · · · · · · · · ·

🔑 **Key Concept Check** What are the three processes that transfer thermal energy?

Density, Thermal Expansion, and Thermal Contraction

In **Figure 12,** the burner transfers thermal energy to the beaker, which transfers thermal energy to the water. Thermal expansion occurs in water nearest the bottom of the beaker. Heating increases the water's volume making it less dense.

At the same time, water molecules at the water's surface transfer thermal energy to the air. This causes cooling and thermal contraction of the water on the surface. The denser water at the surface sinks to the bottom, forcing the less dense water upward. This cycle continues until all the water in the beaker is at the same temperature.

Convection Currents in Earth's Atmosphere

The movement of fluids in a cycle because of convection is a **convection current.** Convection currents circulate the water in Earth's oceans and other bodies of water. They also circulate the air in a room, and the materials in Earth's interior. Convection currents also move matter and thermal energy from inside the Sun to its surface.

On Earth, convection currents move air between the equator and latitudes near 30°N and 30°S. This plays an important role in Earth's climates, as shown in **Figure 13.**

Convection Currents in Earth's Atmosphere

Figure 13 Convection currents in the atmosphere influence the locations of rain forests and deserts.

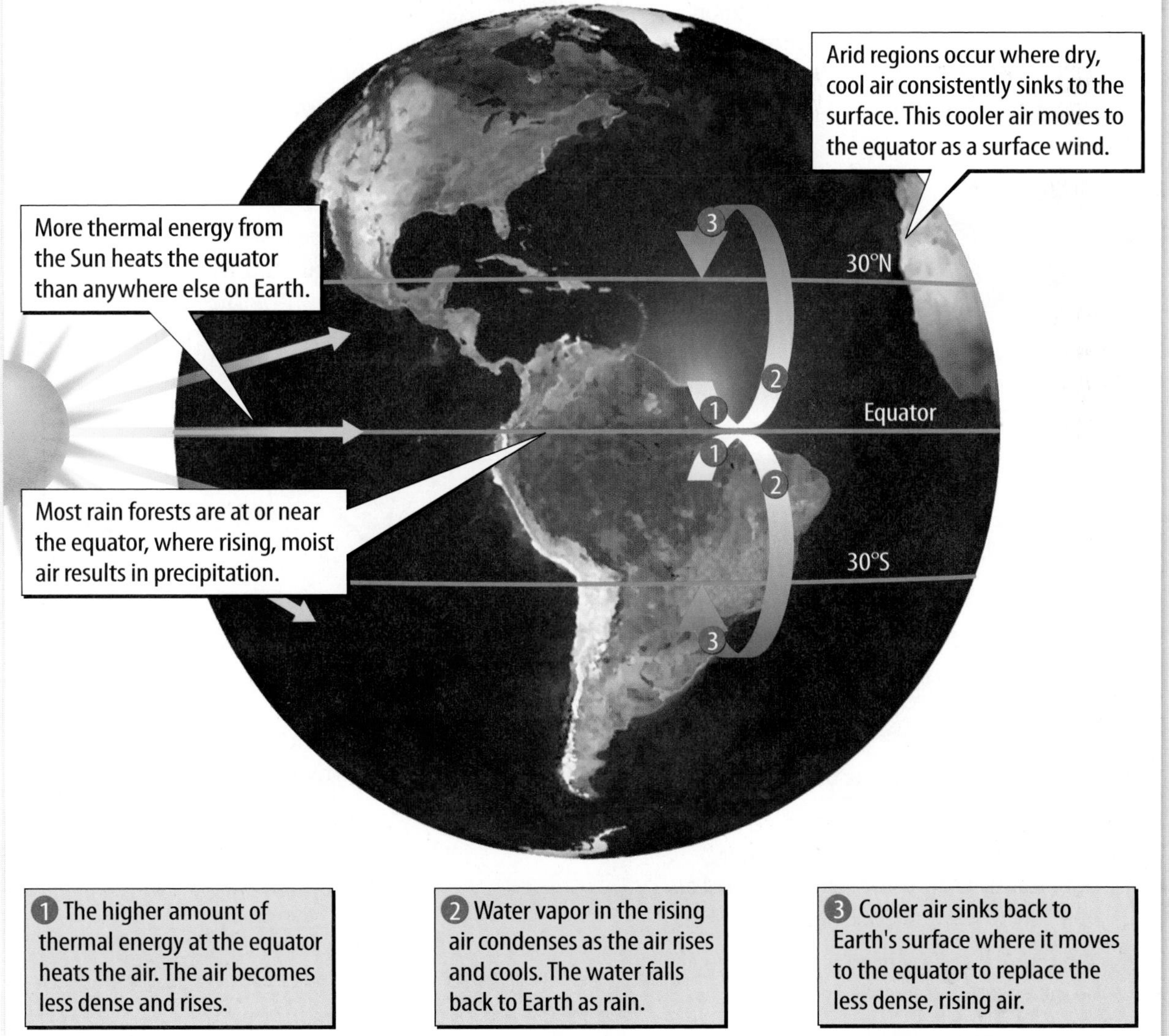

Arid regions occur where dry, cool air consistently sinks to the surface. This cooler air moves to the equator as a surface wind.

More thermal energy from the Sun heats the equator than anywhere else on Earth.

Most rain forests are at or near the equator, where rising, moist air results in precipitation.

30°N

Equator

30°S

❶ The higher amount of thermal energy at the equator heats the air. The air becomes less dense and rises.

❷ Water vapor in the rising air condenses as the air rises and cools. The water falls back to Earth as rain.

❸ Cooler air sinks back to Earth's surface where it moves to the equator to replace the less dense, rising air.

Lesson 2 Review

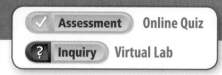
Visual Summary

When a material has a low specific heat, transferring a small amount of energy to the material increases its temperature significantly.

Thermal energy can be transferred through radiation, conduction, or convection.

When a material is heated, the thermal energy of the material increases and the material expands.

FOLDABLES

Use your lesson Foldable to review the lesson. Save your Foldable for the project at the end of the chapter.

What do you think NOW?

You first read the statements below at the beginning of the chapter.

3. It takes a large amount of energy to significantly change the temperature of an object with a low specific heat.

4. The thermal energy of an object can never be increased or decreased.

Did you change your mind about whether you agree or disagree with the statements? Rewrite any false statements to make them true.

Use Vocabulary

1. The transfer of thermal energy by electromagnetic waves is _____.

2. **Define** *convection* in your own words.

Understand Key Concepts

3. **Contrast** radiation with conduction.

4. Why do hot-air balloons rise?
   A. thermal conduction
   B. thermal convection
   C. thermal expansion
   D. thermal radiation

5. **Infer** why the sauce on a hot pizza burns your mouth but the crust of the pizza does not burn your mouth.

Interpret Graphics

6. **Analyze** Two cubes with the same mass and volume are heated in the same pan of water. The graph below shows the change in temperature with time. Which cube has the higher specific heat?

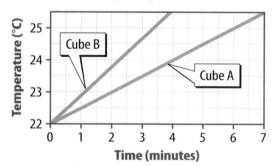

7. **Organize** Copy and fill in the graphic organizer to show how thermal energy is transferred.

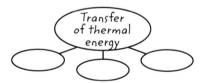

Critical Thinking

8. **Explain** Why do you use a pot holder when taking hot food out of the oven?

Insulating the Home

It's what's between the walls that matters.

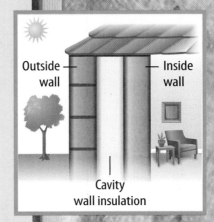

The first requirement of a shelter is to protect you from the weather. If the weather where you live is mild year-round, almost any kind of shelter will do. However, basic shelters, such as huts and tents, are not comfortable during cold winters or hot summers. Over many centuries, societies have experimented with thermal insulators that keep the inside of a shelter warm during winter and cool during summer.

One of the first thermal insulators used in shelters was air. Because air is a poor conductor of thermal energy, using cavity walls became a common form of insulating homes in the United States.

An air gap was not the perfect solution, however. Convection currents in the cavity carried some thermal energy across the gap. At first, no one seemed to mind. But, in the 1970s, the cost of heating and cooling homes suddenly increased. People began looking for a better way to reduce the transfer of thermal energy between the outside and the inside.

To meet the growing demand for better insulation, scientists began researching to find better insulation materials. If you could stop the convection currents, they reasoned, you could stop the transfer of thermal energy. One way was to use materials such as polymer foam or fiberglass. Each of these trapped air between the walls and held it there.

But how do you install insulation if your house is already built? You poke holes in the walls and blow it in! This process has little effect on your home's structure, and it decreases the cost of heating and cooling the house.

Outside wall — Inside wall

Cavity wall insulation

It's Your Turn

MAKE A POSTER A material's insulating ability is rated with an R-value. Find out what an R-value is. Then make a poster showing the ratings of common materials.

Using Thermal Energy

Reading Guide

Key Concepts
ESSENTIAL QUESTIONS

- How does a thermostat work?

- How does a refrigerator keep food cold?

- What are the energy transformations in a car engine?

Vocabulary

heating appliance p. 215

thermostat p. 216

refrigerator p. 216

heat engine p. 218

g Multilingual eGlossary

Inquiry Concentrating Energy?

This power plant uses mirrors to focus light toward a tower. The tower then transforms some of the light into thermal energy. In what ways do we use thermal energy?

How can you transform energy?

If you rub your hands together very quickly, do they become warm? Where does the thermal energy come from?

1. Read and complete a lab safety form.
2. Copy the table into your Science Journal.
3. Place a **thermometer strip** on the surface of a **block of wood.** Record the temperature after the thermometer stops changing color.

	Starting temp (°C)	Ending temp (°C)
30 s		
60 s		

4. Remove the thermometer and rub the wood vigorously with **sandpaper** for 30 seconds. Quickly replace the thermometer, and record the temperature.
5. Repeat steps 3 and 4 on another part of the wood. This time, sand the wood for 60 seconds.

Think About This

1. Did the temperature of the wood change? Why or why not?
2. When did the wood have the highest temperature? Explain this result.
3. 🔑 **Key Concept** What energy transformations take place in this activity?

Thermal Energy Transformations

You can convert other forms of energy into thermal energy. Repeatedly stretching a rubber band makes it hot. Burning wood heats the air. A toaster gets hot when you turn it on.

You also can convert thermal energy into other forms of energy. Burning coal can generate electricity. Thermostats transform thermal energy into mechanical energy that switch heaters on and off. When you convert energy from one form to another, you can use the energy to perform useful tasks.

Remember that energy cannot be created or destroyed. Even though many devices transform energy from one form to another or transfer energy from one place to another, the total amount of energy does not change.

Heating Appliances

A device that converts electric energy into thermal energy is a **heating appliance.** Curling irons, coffeemakers, and clothes irons are some examples of heating appliances.

Other devices, such as computers and cell phones, also become warm when you use them. This is because some electric energy always is converted to thermal energy in an electronic device. However, the thermal energy that most electronic devices generate is not used for any purpose.

FOLDABLES

Make a vertical four-tab book. Label it as shown. Use it to explain the energy transformation that occurs in each device.

Heating Appliances
Heat Engines
Refrigerators
Thermostats

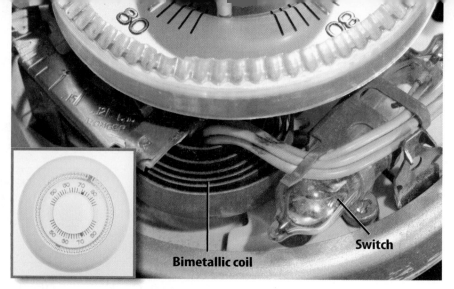

Figure 14 The coil in a thermostat contains two different metals that expand at two different rates.

Switch

Bimetallic coil

Thermostats

You might have heard the furnace in your house or in your classroom turn on in the winter. After the room warms, the furnace turns off. *A* **thermostat** *is a device that regulates the temperature of a system.* Kitchen refrigerators, toasters, and ovens are all equipped with thermostats.

Most thermostats used in home heating systems contain a bimetallic coil. A bimetallic coil is made of two types of metal joined together and bent into a coil, as shown in **Figure 14.** The metal on the inside of the coil expands and contracts more than the metal on the outside of the coil. After the room warms, the thermal energy in the air causes the bimetallic coil to uncurl slightly. This moves a switch that turns off the furnace. As the room cools, the metal on the inside of the coil contracts more than the metal on the outside, curling the coil tighter. This moves the switch in the other direction, turning on the furnace.

Key Concept Check How does the bimetallic coil in a thermostat respond to heating and cooling?

Refrigerators

A device that uses electric energy to transfer thermal energy from a cooler location to a warmer location is called a **refrigerator.** Recall that thermal energy naturally flows from a warmer area to a cooler area. The opposite might seem impossible. But, that is exactly how your refrigerator works. So, how does a refrigerator move thermal energy from its cold inside to the warm air outside? Pipes that surround the refrigerator are filled with a fluid, called a coolant, that flows through the pipes. Thermal energy from inside the refrigerator transfers to the coolant, keeping the inside of the refrigerator cold.

Vaporizing the Coolant

A coolant is a substance that evaporates at a low temperature. In a refrigerator, a coolant is pumped through pipes on the inside and the outside of the refrigerator. The coolant, which begins as a liquid, passes through an expansion valve and cools. As the cold gas flows through pipes inside the refrigerator, it absorbs thermal energy from the refrigerator compartment and vaporizes. The coolant gas becomes warmer, and the inside of the refrigerator becomes cooler.

Condensing the Coolant

The coolant flows to an electric compressor at the bottom of the refrigerator. Here, the coolant is compressed, or forced into a smaller space, which increases its thermal energy. Then, the gas is pumped through condenser coils. In the coils, the thermal energy of the gas is greater than that of the surrounding air. This causes thermal energy to flow from the coolant gas to the air behind the refrigerator. As thermal energy is removed from the gas, it condenses, or becomes liquid. Then, the liquid coolant is pumped up through the expansion valve. The cycle repeats.

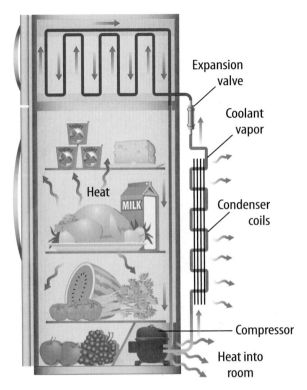

Figure 15 🔑 Coolant in a refrigerator moves thermal energy from inside to outside the refrigerator.

 Key Concept Check How does a refrigerator keep food cold?

Inquiry MiniLab　　　　　　　　　**10 minutes**

Can thermal energy be used to do work? 🥽 🧪 🧤 ✋

You know you can raise the thermal energy of a substance by doing work on it. Is the opposite true? Can thermal energy cause something to move?

1 Read and complete a lab safety form.

2 Add 10 mL of water to a **100-mL beaker.**

3 Place a **small square of aluminum foil** over the top of the beaker.

4 Place the beaker on a **hot plate,** and turn it on. Observe the results and record them in your Science Journal.

Analyze and Conclude

1. **Infer** Is thermal energy used to do work in this lab? Explain your answer.

2. 🔑 **Key Concept** Is thermal energy transformed into another form of energy in this experiment? If so, what is the other form of energy?

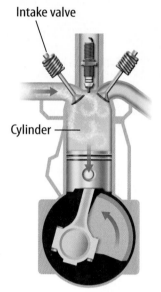
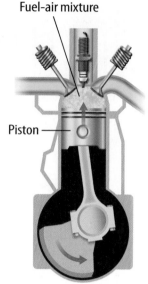
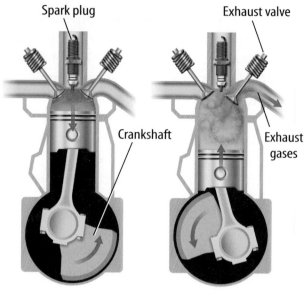

Intake valve

Fuel-air mixture

Spark plug

Exhaust valve

Cylinder

Piston

Crankshaft

Exhaust gases

❶ The intake valve opens as the piston moves downward, drawing a mixture of gasoline and air into the cylinder.

❷ The intake valve closes as the piston moves upward, compressing the fuel-air mixture.

❸ A spark plug ignites the fuel-air mixture. As the mixture burns, hot gases expand, pushing the piston down.

❹ As the piston moves up, the exhaust valve opens, and the hot gases are pushed out of the cylinder.

Figure 16 Internal combustion engines transform the chemical energy from fuel to thermal energy, which then produces mechanical energy.

Heat Engines

A typical automobile engine is a heat engine. *A* **heat engine** *is a machine that converts thermal energy into mechanical energy.* When a heat engine converts thermal energy into mechanical energy, the mechanical energy moves the vehicle. Most cars, buses, boats, trucks, and lawn mowers use a type of heat engine called an internal combustion engine. **Figure 16** shows how one type of internal combustion engine converts thermal energy into mechanical energy.

Perhaps you have heard someone refer to a car as having a six-cylinder engine. A cylinder is a tube with a piston that moves up and down. At one end of the cylinder a spark ignites a fuel-air mixture. The ignited fuel-air mixture expands and pushes the piston down. This action occurs because the fuel's chemical energy converts to thermal energy. Some of the thermal energy immediately converts to mechanical energy.

A heat engine is not efficient. Most automobile engines only convert about 20 percent of the chemical energy in gasoline into mechanical energy. The remaining energy from the gasoline is lost to the environment.

Key Concept Check What is one form of energy that is output from a heat engine?

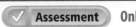

Visual Summary

A bimetallic coil inside a thermostat controls a switch that turns a heating or cooling device on or off.

A refrigerator keeps food cold by moving thermal energy from the inside of the refrigerator out to the refrigerator's surroundings.

In a car engine, chemical energy in fuel is transformed into thermal energy. Some of this thermal energy is then transformed into mechanical energy.

FOLDABLES®

Use your lesson Foldable to review the lesson. Save your Foldable for the project at the end of the chapter.

What do you think NOW?

You first read the statements below at the beginning of the chapter.

5. Car engines create energy.

6. Refrigerators cool food by moving thermal energy from inside the refrigerator to the outside.

Did you change your mind about whether you agree or disagree with the statements? Rewrite any false statements to make them true.

Use Vocabulary

❶ A _____ is a device that converts electric energy into thermal energy.

❷ **Explain** how an internal combustion engine works.

Understand Key Concepts

❸ **Describe** the path of thermal energy in a refrigerator.

❹ Which sequence describes the energy transformation in an automobile engine?
   A. chemical→thermal→mechanical
   B. thermal→kinetic→potential
   C. thermal→mechanical→potential
   D. thermal→chemical→mechanical

❺ **Explain** how a thermostat uses electric energy, mechanical energy, and thermal energy.

Interpret Graphics

❻ **Predict** Suppose you pointed a hair dryer at the device pictured below and turned on the hair dryer. What would happen?

❼ **Sequence** Copy the graphic organizer below. Use it to show the steps involved in one cycle of an internal combustion engine.

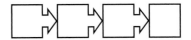

Critical Thinking

❽ **Explain** how two of the devices you read about in this chapter could be used in one appliance.

Design an Insulated Container

Materials

aluminum foil

self-sealing plastic bag

triple-beam balance

creative building materials

office supplies

Also Needed

frozen fruit pop, foam packing peanuts, rubber bands

Safety

Many refrigerated or frozen food products must be kept cold as they are transported long distances. Meat or fresh fruits might travel from South America to grocery stores in the United States. Imagine that you have been hired to design a container that will keep a frozen fruit pop from melting for as long as possible.

Ask a Question

How can you construct a container that will prevent a frozen fruit pop inside a plastic bag from melting? Think about thermal energy transfer by conduction, convection, and radiation. You will begin with a shoe box, but you can modify it in any way. Consider the materials you have available. Ask yourself what material you can bring from home that might slow the melting of a frozen fruit pop.

Make Observations

1. Read and complete a lab safety form.
2. In your Science Journal, write your ideas about
   - how you can you reduce the amount of thermal energy moving by conduction, convection, and radiation;
   - what materials you will use inside and outside your box;
   - what materials you will need to bring from home.
3. Outline the steps in preparing for your box. Have your teacher check your procedures. Decide who will obtain which materials before the next lab period. Design a logo for your container.
4. As a class, decide how many hours you will wait before checking the condition of your frozen fruit pop.

Form a Hypothesis

5. Formulate a hypothesis explaining why the materials you use inside your bag will be effective in insulating the frozen fruit pop. Remember, your hypothesis should be a testable explanation based on observations.

Test Your Hypothesis

6 On the second lab day, follow the steps you have outlined and prepare your container. Check it over one more time to be sure you have accounted for all ways that thermal energy could enter or leave the box.

7 Obtain a frozen fruit pop. Place it inside a self-sealing plastic bag. Seal the bag. Quickly measure and record its mass. Attach your logo and return the pop to the freezer.

8 On the third lab day, remove your frozen fruit pop from the freezer. Do not open the plastic bag. Place your frozen fruit pop in your container and seal it. Place your container in a location assigned by your teacher.

9 After the set amount of time, remove the fruit pop from the container. Open the plastic bag, and pour off any melted juice. Reseal the bag. Measure and record the mass.

Analyze and Conclude

10 **Calculate** What percentage of your fruit pop remained frozen? How long do you think it would take for the fruit pop to completely melt in your container? Justify your answer.

11 **Analyze** What are some possible ways thermal energy entered your bag? How could you improve the package on another try?

12 🔵 **The Big Idea** How would you modify your design to keep something hot inside the bag? Explain your answer.

Communicate Your Results

Make a class graph showing the percentages of the different frozen fruit pops remaining. Discuss why some packages were more or less effective.

Explore designs for portable coolers. What are the most effective portable packages that keep things hot or cold without external cooling or heating?

Lab Tips

☑ Keep in mind that you are trying to keep thermal energy out of the package.

☑ The length of the test time you decide on should be long enough to allow some of the fruit pop to melt.

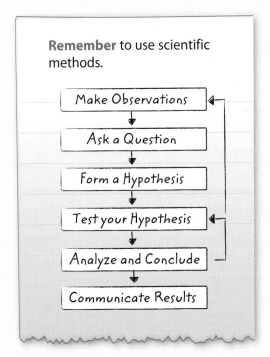

Remember to use scientific methods.

Make Observations
↓
Ask a Question
↓
Form a Hypothesis
↓
Test your Hypothesis
↓
Analyze and Conclude
↓
Communicate Results

Chapter 6 Study Guide

Thermal energy can be transferred by conduction, radiation, and convection. Thermal energy also can be transformed into other forms of energy and used in devices such as thermostats, refrigerators, and automobile engines.

Key Concepts Summary 🔑

Vocabulary

Lesson 1: Thermal Energy, Temperature, and Heat

- The **temperature** of a material is the average kinetic energy of the particles that make up the material.

- **Heat** is the movement of **thermal energy** from a material or area with a higher temperature to a material or area with a lower temperature.

- When a material is heated, the material's temperature changes.

thermal energy p. 198
temperature p. 199
heat p. 201

Lesson 2: Thermal Energy Transfers

- When a material has a low **specific heat**, transferring a small amount of energy to the material increases its temperature significantly.

- When a material is heated, the thermal energy of the material increases and the material expands.

- Thermal energy can be transferred by **conduction, radiation,** or **convection.**

radiation p. 205
conduction p. 206
thermal conductor p. 206
thermal insulator p. 206
specific heat p. 207
thermal contraction p. 208
thermal expansion p. 208
convection p. 210
convection current p. 211

Lesson 3: Using Thermal Energy

- The two different metals in a bimetallic coil inside a **thermostat** expand and contract at different rates. The bimetallic coil curls and uncurls, depending on the thermal energy of the air, pushing a switch that turns a heating or cooling device on or off.

- A **refrigerator** keeps food cold by moving thermal energy from inside the refrigerator out to the refrigerator's surroundings.

- In a car engine, chemical energy in fuel is transformed into thermal energy. Some of this thermal energy is then transformed into mechanical energy.

heating appliance p. 215
thermostat p. 216
refrigerator p. 216
heat engine p. 218

[=] Review
• Personal Tutor
• Vocabulary eGames
• Vocabulary eFlashcards

FOLDABLES® Chapter Project

Assemble your lesson Foldables as shown to make a Chapter Project. Use the project to review what you have learned in this chapter.

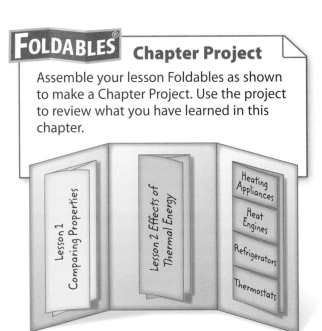

Lesson 1 Comparing Properties

Lesson 2 Effects of Thermal Energy

Heating Appliances

Heat Engines

Refrigerators

Thermostats

Use Vocabulary

1. When you increase the _____ of a cup of hot cocoa, you increase the average kinetic energy of the particles that make up the hot cocoa.

2. The increase in volume of a material when heated is _____.

3. A(n) _____ is used to control the temperature in a room.

4. Thermal energy is transferred by _____ between two objects that are touching.

5. A fluid moving in a circular pattern because of convection is a _____.

6. Define *heating appliance* in your own words.

Link Vocabulary and Key Concepts

((O)) Concepts in Motion   Interactive Concept Map

Copy this concept map, and then use vocabulary terms from the previous page to complete the concept map.

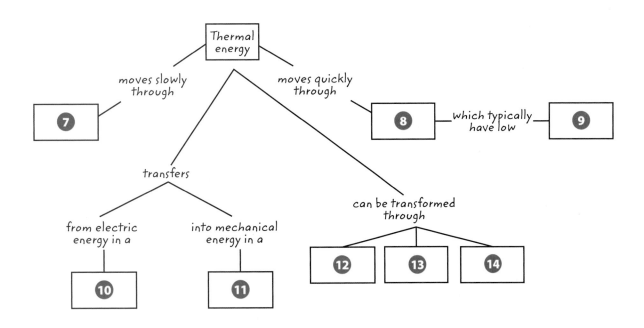

Chapter 6 Review

Understand Key Concepts

1 Which would decrease a material's thermal energy?
- A. heating the material
- B. increasing the kinetic energy of the particles that make up the material
- C. increasing the temperature of the material
- D. moving the material to a location where the temperature is lower

2 You put a metal spoon in a bowl of hot soup. Why does the spoon feel hotter than the outside of the bowl?
- A. The bowl is a better conductor than the spoon.
- B. The bowl has a lower specific heat than the spoon.
- C. The spoon is a good thermal insulator.
- D. The spoon transfers thermal energy better than the bowl does.

3 In the picture to the right, thermal energy moves from the
- A. glass to the air.
- B. lemonade to the air.
- C. ice to the lemonade.
- D. air to the lemonade.

4 Which has the lowest specific heat?
- A. an object that is made out of metal
- B. an object that does not transfer thermal energy easily
- C. an object with electrons that do not move easily
- D. an object that requires a lot of energy to change its temperature

5 Which does NOT occur in an internal combustion engine?
- A. Most of the thermal energy is wasted.
- B. Thermal energy forces the piston downward.
- C. Thermal energy is converted into chemical energy.
- D. Thermal energy is converted into mechanical energy.

6 Which statement about radiation is correct?
- A. In solids, radiation transfers electromagnetic energy, but not thermal energy.
- B. Cooler objects radiate the same amount of thermal energy as warmer objects.
- C. Radiation occurs in fluids such as gas and water, but not solids such as metals.
- D. Radiation transfers thermal energy from the Sun to Earth.

7 The device below detects an increase in room temperature as

- A. an increase in thermal energy causes a bimetallic coil to curl.
- B. an increase in thermal energy causes a bimetallic coil to uncurl.
- C. a switch causes a bimetallic coil to curl.
- D. a switch causes a bimetallic coil to uncurl

8 Which is the lowest temperature?
- A. 0°C
- B. 0°F
- C. 32°F
- D. 273 K

9 Which energy conversion typically occurs in a heating appliance?
- A. chemical energy to thermal energy
- B. electric energy to thermal energy
- C. thermal energy to chemical energy
- D. thermal energy to mechanical energy

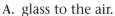

Critical Thinking

10 **Compare** A swimming pool with a temperature of 30°C has more thermal energy than a cup of soup with a temperature of 60°C. Explain why this is so.

11 **Contrast** A spoon made of aluminum and a spoon made of steel have the same mass. The aluminum spoon has a higher specific heat than the steel spoon. Which spoon becomes hotter when placed in a pan of boiling water?

12 **Describe** How do convection currents influence Earth's climate?

13 **Diagram** A room has a heater on one side and an open window letting in cool air on the opposite side. Diagram the convection current in the room. Label the warm air and the cool air.

14 **Evaluate** When engineers build bridges, they separate sections of the roadway with expansion joints such as the one below that allow movement between the sections. Why are expansion joints necessary?

15 **Explain** Why is conduction slower in a gas than in a liquid or a solid?

Writing in Science

16 **Research** various types of heat engines that have been developed throughout history. Write 3–5 paragraphs explaining the energy transformations in one of these engines.

REVIEW THE BIG IDEA

17 **Describe** each of the three ways thermal energy can be transferred. Give an example of each.

18 What do the different colors in this photograph indicate?

Math Skills ×÷ ⊟ Review

── Math Practice ──

Convert Between Temperature Scales

19 If water in a bath is at 104°F, then what is the temperature of the water in degrees Celsius?

20 Convert −40°C to degrees Fahrenheit.

Standardized Test Practice

Record your answers on the answer sheet provided by your teacher or on a sheet of paper.

Multiple Choice

1 Which statement describes the thermal energy of an object?

  A kinetic energy of particles + potential energy of particles

  B kinetic energy of particles ÷ number of particles

  C potential energy of particles ÷ number of particles

  D kinetic energy of particles ÷ (kinetic energy of particles + potential energy of particles)

2 Which term describes a transfer of thermal energy?

  A heat

  B specific heat

  C temperature

  D thermal energy

Use the figures below to answer question 3.

Sample X   **Sample Y**

3 The figures show two different samples of air. In what way do they differ?

  A Sample X is at a higher temperature than sample Y.

  B Sample X has a higher specific heat than sample Y.

  C Particles of sample Y have a higher average kinetic energy than those of sample X.

  D Particles of sample Y have a higher average thermal energy than those of sample X.

Use the table below to answer question 4.

Material	Specific Heat (in J/g·K)
Air	1.0
Copper	0.4
Water	4.2
Wax	2.5

4 The table shows the specific heat of four materials. Which statement can be concluded from the information in the table?

  A Copper is a thermal insulator.

  B Wax is a thermal conductor.

  C Air takes the most thermal energy to change its temperature.

  D Water takes the most thermal energy to change its temperature.

5 Which term describes what happens to a cold balloon when placed in a hot car?

  A thermal conduction

  B thermal contraction

  C thermal expansion

  D thermal insulation

6 A girl stirs soup with a metal spoon. Which process causes her hand to get warmer?

  A conduction

  B convection

  C insulation

  D radiation

7 In a thermostat's coil, what causes the two metals in the strip to curl and uncurl?

  A They contract at the same rate when cooled.

  B They expand at different rates when heated.

  C They have the same specific heat.

  D They melt at different temperatures.

Use the figure to below to answer questions 8–10.

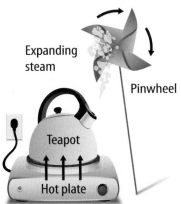

8 Which term describes the transfer of thermal energy between the hot plate and the teapot?

  A conduction

  B convection

  C insulation

  D radiation

9 Which energy transformations are taking place in this system?

  A electrical → thermal → chemical

  B electrical → thermal → mechanical

  C thermal → electrical → chemical

  D thermal → electrical → mechanical

10 What kind of machine is represented by the hot plate, the teapot, the steam, and the pinwheel working together?

  A bimetallic coil

  B heat engine

  C refrigerator

  D thermostat

Constructed Response

Use the figure to answer questions 11 and 12.

11 The foam cooler and the metal pan both contain ice. Describe the energy transfers that cause the ice to melt in each container.

12 After 1 hour, the ice in the metal pan had melted more than the ice in the foam cooler. What is it about the containers that could explain the difference in the melting rates?

13 What causes the air around a refrigerator to become warmer as the refrigerator is cooling the air inside it?

14 How does a car's internal combustion engine convert thermal energy to mechanical energy?

NEED EXTRA HELP?														
If You Missed Question...	1	2	3	4	5	6	7	8	9	10	11	12	13	14
Go to Lesson...	1	1	1	2	2	2	3	2	3	3	2	2	3	3

Foundations of Chemistry

THE BIG IDEA What is matter, and how does it change?

Inquiry Why does it glow?

This siphonophore (si FAW nuh fawr) lives in the Arctic Ocean. Its tentacles have a very powerful sting. However, the most obvious characteristic of this organism is the way it glows.

- What might cause the siphonophore to glow?

- How do you think its glow helps the siphonophore survive?

- What changes happen in the matter that makes up the organism?

Get Ready to Read

What do you think?

Before you read, decide if you agree or disagree with each of these statements. As you read this chapter, see if you change your mind about any of the statements.

1. The atoms in all objects are the same.

2. You cannot always tell by an object's appearance whether it is made of more than one type of atom.

3. The weight of a material never changes, regardless of where it is.

4. Boiling is one method used to separate parts of a mixture.

5. Heating a material decreases the energy of its particles.

6. When you stir sugar into water, the sugar and water evenly mix.

7. When wood burns, new materials form.

8. Temperature can affect the rate at which chemical changes occur.

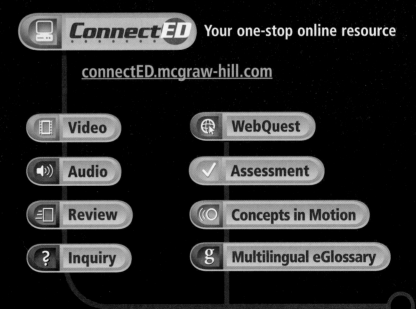

Connect ED Your one-stop online resource

connectED.mcgraw-hill.com

- Video
- WebQuest
- Audio
- Assessment
- Review
- Concepts in Motion
- Inquiry
- Multilingual eGlossary

Classifying Matter

Reading Guide

Key Concepts
ESSENTIAL QUESTIONS

- What is a substance?
- How do atoms of different elements differ?
- How do mixtures differ from substances?
- How can you classify matter?

Vocabulary

matter p. 231

atom p. 231

substance p. 233

element p. 233

compound p. 234

mixture p. 235

heterogeneous mixture p. 235

homogeneous mixture p. 235

dissolve p. 235

g Multilingual eGlossary

** Video**

- BrainPOP®
- Science Video
- What's Science Got to do With It?

Inquiry Making Green?

You probably have mixed paints together. Maybe you wanted green paint and had only yellow paint and blue paint. Perhaps you watched an artist mixing several tints get the color he or she needed. In all these instances, the final color came from mixing colors together and not from changing the color of a paint.

How do you classify matter?

An object made of paper bound together might be classified as a book. Pointed metal objects might be classified as nails or needles. How can you classify an item based on its description?

1. Read and complete a lab safety form.

2. Place the **objects** on a table. Discuss how you might separate the objects into groups with these characteristics:

   a. Every object is the same and has only one part.
   b. Every object is the same but is made of more than one part.
   c. Individual objects are different. Some have one part, and others have more than one part.

3. Identify the objects that meet the requirements for group *a,* and record them in your Science Journal. Repeat with groups *b* and *c.* Any object can be in more than one group.

Think About This

1. Does any object from the bag belong in all three of the groups (*a, b,* and *c*)? Explain.

2. What objects in your classroom would fit into group *b?*

3. **Key Concept** What descriptions would you use to classify items around you?

Understanding Matter

Have you ever seen a rock like the one in **Figure 1?** Why are different parts of the rock different in color? Why might some parts of the rock feel harder than other parts? The parts of the rock look and feel different because they are made of different types of matter. **Matter** is *anything that has mass and takes up space.* If you look around, you will see many types of matter. If you are in a classroom, you might see things made of metal, wood, or plastic. If you go to a park, you might see trees, soil, or water in a pond. If you look up at the sky, you might see clouds and the Sun. All of these things are made of matter.

Everything you can see is matter. However, some things you cannot see also are matter. Air, for example, is matter because it has mass and takes up space. Sound and light are not matter. Forces and energy also are not matter. To decide whether something is matter, ask yourself if it has mass and takes up space.

An **atom** *is a small particle that is a building block of matter.* In this lesson, you will explore the parts of an atom and read how atoms can differ. You will also read how different arrangements of atoms make up the many types of matter.

WORD ORIGIN

matter
from Latin *materia,* meaning "material, stuff"

Figure 1 You can see different types of matter in this rock.

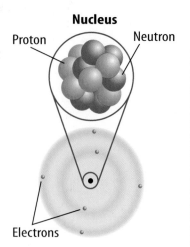

Nucleus

Proton

Neutron

Electrons

Figure 2 An atom has electrons moving in an area outside a nucleus. Protons and neutrons make up the nucleus.

 Review **Personal Tutor**

Atoms

To understand why there are so many types of matter, it helps if you first learn about the parts of an atom. Look at the diagram of an atom in **Figure 2.** At the center of an atom is a nucleus. Protons, which have a positive charge, and neutrons, which have a neutral charge, make up the nucleus. Negatively charged particles, or electrons, move quickly throughout an area around the nucleus called the electron cloud.

✓ **Reading Check** What are the parts of an atom?

Not all atoms have the same number of protons, neutrons, and electrons. Atoms that have different numbers of protons differ in their properties. You will read more about the differences in atoms on the next page.

An atom is almost too small to imagine. Think about how thin a human hair is. The diameter of a human hair is about a million times greater than the diameter of an atom. In addition, an atom is about 10,000 times wider than its nucleus! Even though atoms are so tiny, they determine the properties of the matter they compose.

Inquiry MiniLab

20 minutes

How can you model an atom?

How can you model an atom out of its three basic parts?

1. Read and complete a lab safety form.

2. Twist the ends of a piece of **florist wire** together to form a ring. Attach two **wires** across the ring to form an *X*.

3. Use **double-sided tape** to join the **large pom-poms** (protons and neutrons), forming a nucleus. Hang the nucleus from the center of the *X* with **fishing line.**

4. Use fishing line to suspend each **small pom-pom** (electron) from the ring so they surround the nucleus.

5. Suspend your model as instructed by your teacher.

Analyze and Conclude

1. **Infer** Based on your model, what can you infer about the relative sizes of protons, neutrons, and electrons?

2. **Model** Why is it difficult to model the location of electrons?

3. 🔑 **Key Concept** Compare your atom with those of other groups. How do they differ?

Substances

You can see that atoms make up most of the matter on Earth. Atoms can combine and arrange in millions of different ways. In fact, these different combinations and arrangements of atoms are what makes up the various types of matter. There are two main classifications of matter—substances and mixtures.

A **substance** *is matter with a composition that is always the same.* This means that a given substance is always make up of the same combination(s) of atoms. Aluminum, oxygen, water, and sugar are examples of substances. Any sample of aluminum is always made up of the same type of atoms, just as samples of oxygen, sugar, and water each are always made of the same combinations of atoms. To gain a better understanding of what makes up substances, let's take a look at the two types of substances—elements and compounds.

 Key Concept Check What is a substance?

Elements

Look at the periodic table of elements on the inside back cover of this book. The substances oxygen and aluminum are on the table. They are both elements. *An* **element** *is a substance that consists of just one type of atom.* Because there are about 115 known elements, there are about 115 different types of atoms. Each type of atom contains a different number of protons in its nucleus. For example, each aluminum atom has 13 protons in its nucleus. The number of protons in an atom is the atomic number of the element. Therefore, the atomic number of aluminum is 13, as shown in **Figure 3.**

The atoms of most elements exist as individual atoms. For example, a roll of pure aluminum foil consists of trillions of individual aluminum atoms. However, the atoms of some elements usually exist in groups. For example, the oxygen atoms in air exist in pairs. Whether the atoms of an element exist individually or in groups, each element contains only one type of atom. Therefore, its composition is always the same.

 Key Concept Check How do atoms of different elements differ?

Figure 3 🔑 Each element on the periodic table consists of just one type of atom.

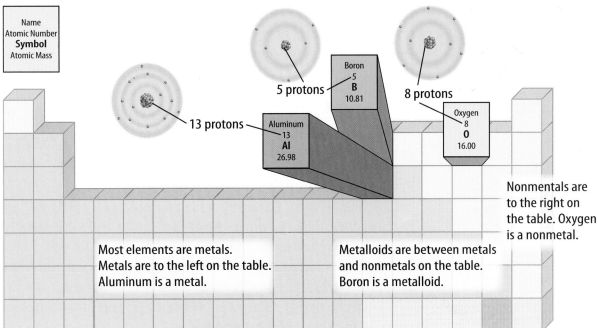

Name
Atomic Number
Symbol
Atomic Mass

5 protons

Boron
5
B
10.81

8 protons

13 protons

Aluminum
13
Al
26.98

Oxygen
8
O
16.00

Nonmentals are to the right on the table. Oxygen is a nonmetal.

Most elements are metals. Metals are to the left on the table. Aluminum is a metal.

Metalloids are between metals and nonmetals on the table. Boron is a metalloid.

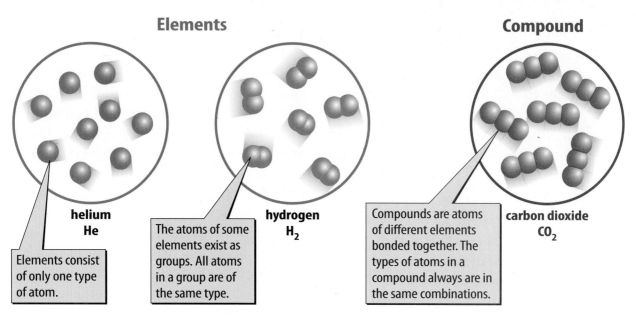

Elements   Compound

helium
He

Elements consist of only one type of atom.

The atoms of some elements exist as groups. All atoms in a group are of the same type.

hydrogen
H₂

Compounds are atoms of different elements bonded together. The types of atoms in a compound always are in the same combinations.

carbon dioxide
CO₂

▲ **Figure 4** 🔑 If a substance contains only one type of atom, it is an element. If it contains more than one type of atom, it is a compound.

Figure 5 Carbon dioxide is a compound composed of carbon and oxygen atoms. ▼

 **Review** Personal Tutor

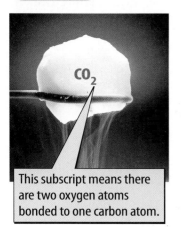

This subscript means there are two oxygen atoms bonded to one carbon atom.

ACADEMIC VOCABULARY

unique
(adjective) having nothing else like it

Compounds

Water is a substance, but it is not an element. It is a compound. *A* **compound** *is a type of substance containing atoms of two or more different elements chemically bonded together.* As shown in **Figure 4,** carbon dioxide (CO_2) is also a compound. It consists of atoms of two different elements, carbon (C) and oxygen (O), bonded together. Carbon dioxide is a substance because the C and the O atoms are always combined in the same way.

Chemical Formulas The combination of symbols and numbers that represents a compound is called a chemical formula. Chemical formulas show the different atoms that make up a compound, using their element symbols. Chemical formulas also help explain how the atoms combine. As illustrated in **Figure 5,** CO_2 is the chemical formula for carbon dioxide. The formula shows that carbon dioxide is made of C and O atoms. The small *2* is called a subscript. It means that two oxygen atoms and one carbon atom form carbon dioxide. If no subscript is written after a symbol, one atom of that element is present in the chemical formula.

Properties of Compounds Think again about the elements carbon and oxygen. Carbon is a black solid, and oxygen is a gas that enables fuels to burn. However, when they chemically combine, they form the compound carbon dioxide, which is a gas used to extinguish fires. A compound often has different properties from the individual elements that compose it. Compounds, like elements, are substances, and all substances have their own **unique** properties.

Mixtures

Another classification of matter is mixtures. *A **mixture** is matter that can vary in composition.* Mixtures are combinations of two or more substances that are physically blended together. The amounts of the substances can vary in different parts of a mixture and from mixture to mixture. Think about sand mixed with water at the beach. The sand and the water do not bond together. Instead, they form a mixture. The substances in a mixture do not combine chemically. Therefore, they can be separated by physical methods, such as filtering.

Heterogeneous Mixtures

Mixtures can differ depending on how well the substances that make them up are mixed. Sand and water at the beach form a mixture, but the sand is not evenly mixed throughout the water. Therefore, sand and water form a heterogeneous mixture. *A **heterogeneous mixture** is a type of mixture in which the individual substances are not evenly mixed.* Because the substances in a heterogeneous mixture are not evenly mixed, two samples of the same mixture can have different amounts of the substances, as shown in **Figure 6.** For example, if you fill two buckets with sand and water at the beach, one bucket might have more sand in it than the other.

Homogeneous Mixtures

Unlike a mixture of water and sand, the substances in mixtures such as apple juice, air, or salt water are evenly mixed. *A **homogeneous mixture** is a type of mixture in which the individual substances are evenly mixed.* In a homogeneous mixture, the particles of individual substances are so small and well-mixed that they are not visible, even with most high-powered microscopes.

A homogeneous mixture also is known as a solution. In a solution, the substance present in the largest amount is called the solvent. All other substances in a solution are called solutes. The solutes dissolve in the solvent. *To **dissolve** means to form a solution by mixing evenly.* Because the substances in a solution, or homogeneous mixture, are evenly mixed, two samples from a solution will have the same amounts of each substance. For example, imagine pouring two glasses of apple juice from the same container. Each glass will contain the same substances (water, sugar, and other substances) in the same amounts. However, because apple juice is a mixture, the amounts of the substances from one container of apple juice to another might vary.

 Key Concept Check How do mixtures differ from substances?

Figure 6 Types of mixtures differ in how evenly their substances are mixed.

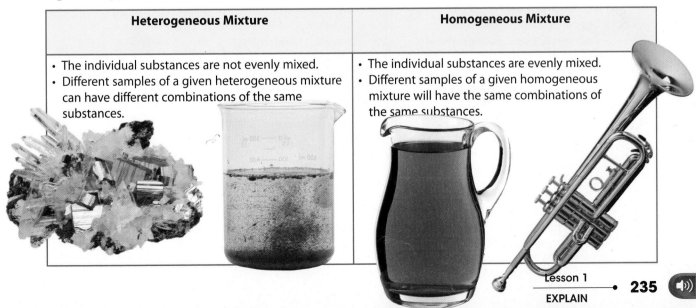

Heterogeneous Mixture	Homogeneous Mixture
• The individual substances are not evenly mixed. • Different samples of a given heterogeneous mixture can have different combinations of the same substances.	• The individual substances are evenly mixed. • Different samples of a given homogeneous mixture will have the same combinations of the same substances.

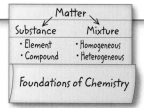

Compounds v. Solutions

If you have a glass of pure water and a glass of salt water, can you tell which is which just by looking at them? You cannot. Both the compound (water) and the solution (salt water) appear identical. How do compounds and solutions differ?

Because water is a compound, its composition does not vary. Pure water is always made up of the same atoms in the same combinations. Therefore, a chemical formula can be used to describe the atoms that make up water (H_2O). Salt water is a homogeneous mixture, or solution. The solute (NaCl) and the solvent (H_2O) are evenly mixed but are not bonded together. Adding more salt or more water only changes the relative amounts of the substances. In other words, the composition varies. Because composition can vary in a mixture, a chemical formula cannot be used to describe mixtures.

Summarizing Matter

You have read in this lesson about classifying matter by the arrangement of its atoms. **Figure 7** is a summary of this classification system.

Key Concept Check How can you classify matter?

Figure 7 Scientists classify matter according to the arrangement of the atoms that make up the matter.

Classifying Matter

Matter
• Anything that has mass and takes up space
• Matter on Earth is made up of atoms.
• Two classifications of matter: substances and mixtures

Substances
• Matter with a composition that is always the same
• Two types of substances: elements and compounds

Element
• Consists of just one type of atom
• Organized on the periodic table
• Each element has a chemical symbol.

Compound
• Two or more types of atoms bonded together
• Properties are different from the properties of the elements that make it up
• Each compound has a chemical formula.

Substances physically combine to form mixtures.

Mixtures can be separated into substances by physical methods.

Mixtures
• Matter that can vary in composition
• Substances are not bonded together.
• Two types of mixtures: heterogeneous and homogeneous

Heterogeneous Mixture
• Two or more substances unevenly mixed
• Different substances are visible by an unaided eye or a microscope.

Homogeneous Mixture—Solution
• Two or more substances evenly mixed
• Different substances cannot be seen even by a microscope.

Lesson 1 Review

Visual Summary

A substance has the same composition throughout. A substance is either an element or a compound.

An atom is the smallest part of an element that has its properties. Atoms contain protons, neutrons, and electrons.

The substances in a mixture are not chemically combined. Mixtures can be either heterogeneous or homogeneous.

FOLDABLES

Use your lesson Foldable to review the lesson. Save your Foldable for the project at the end of the chapter.

What do you think NOW?

You first read the statements below at the beginning of the chapter.

1. The atoms in all objects are the same.

2. You cannot always tell by an object's appearance whether it is made of more than one type of atom.

Did you change your mind about whether you agree or disagree with the statements? Rewrite any false statements to make them true.

Use Vocabulary

1 Substances and mixtures are two types of _____.

2 **Use the term** *atom* in a complete sentence.

3 **Define** *dissolve* in your own words.

Understand Key Concepts

4 **Explain** why aluminum is a substance.

5 The number of _____ always differs in atoms of different elements.
- **A.** electrons
- **C.** neutrons
- **B.** protons
- **D.** nuclei

6 **Distinguish** between a heterogeneous mixture and a homogeneous mixture.

7 **Classify** Which term describes matter that is a substance made of different kinds of atoms bonded together?

Interpret Graphics

8 **Describe** what each letter and number means in the chemical formula below.

$$C_6H_{12}O_6$$

9 **Organize Information** Copy and fill in the graphic organizer below to classify matter by the arrangement of its atoms.

Type of Matter	Description

Critical Thinking

10 **Reorder** the elements aluminum, oxygen, fluorine, calcium, and hydrogen from the least to the greatest number of protons. Use the periodic table if needed.

11 **Evaluate** this statement: Substances are made of two or more types of elements.

HOW IT WORKS

U.S. Mint

How Coins are Made

In 1793, the U.S. Mint produced more than 11,000 copper pennies and put them into circulation. Soon after, gold and silver coins were introduced as well. Early pennies were made of 95 percent copper and 5 percent zinc. Today's penny contains much more zinc than copper and is much less expensive to produce. Quarters, dimes, and nickels, once made of silver, are now made of copper-nickel alloy.

Cu 2.5%
Zn 97.5%

Ni 8.3%
Cu 91.7%

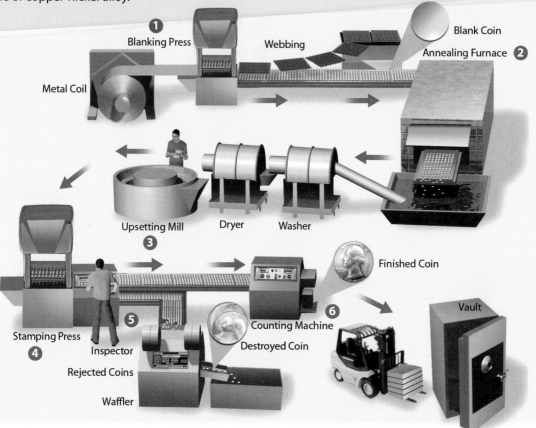

Blanking Press
Webbing
Blank Coin
Annealing Furnace ❷
Metal Coil
❶

Upsetting Mill
Dryer
Washer
❸

Stamping Press
❹
Inspector
Rejected Coins
Waffler
❺
Counting Machine
Destroyed Coin
❻
Finished Coin
Vault

How Coins Are Made

❶ **Blanking** For nickels, dimes, quarters, half-dollars, and coin dollars, a strip of 13-inch-wide metal is fed through a blanking press, which punches out round discs called blanks. The leftover webbing strip is saved for recycling. Ready-made blanks are purchased for making the penny.

❷ **Annealing, Washing, Drying** Blanks are softened in an annealing furnace, which makes the metal less brittle. The blanks are then run through a washer and a dryer.

❸ **Upsetting** Usable blanks are put through an upsetting mill, which creates a rim around the edges of each blank.

❹ **Striking** The blanks then go to the stamping press, where they are imprinted with designs and inscriptions.

❺ **Inspection** Once blanks leave the stamping press, inspectors check a few coins from each batch. Coins that are defective go to the waffler in preparation for recycling.

❻ **Counting and Bagging** A machine counts the finished coins then drops them into large bags that are sealed shut. The coins are then taken to storage before being shipped to Federal Reserve Banks and then to your local bank.

It's Your Turn

COMPARE Collect a variety of coins that includes both older and current coins. Observe and compare their properties. Using the dates of the coins' production, utilize library or Internet sources to research the composition of metals used.

Lesson 2

Physical Properties

Reading Guide

Key Concepts

ESSENTIAL QUESTIONS

- What are some physical properties of matter?
- How are physical properties used to separate mixtures?

Vocabulary

physical property p. 240

mass p. 242

density p. 243

solubility p. 244

 Multilingual eGlossary

Video   **Science Video**

Inquiry Panning by Properties?

The man lowers his pan into the waters of an Alaskan river and scoops up a mixture of water, sediment, and hopefully gold. As he moves the pan in a circle, water sloshes out of it. If he is careful, gold will remain in the pan after the water and sediment are gone. What properties of water, sediment, and gold enable this man to separate this mixture?

Launch Lab

15 minutes

Can you follow the clues?

Clues are bits of information that help you solve a mystery. In this activity, you will use clues to help identify an object in the classroom.

1. Read and complete a lab safety form.

2. Select one **object** in the room. Write a different clue about the object on each of five **index cards.** Clues might include one or two words that describe the object's color, size, texture, shape, or any property you can observe with your senses.

3. Stack your cards face down. Have your partner turn over one card and try to identify the object. Respond either "yes" or "no."

4. Continue turning over cards until your partner identifies your object or runs out of cards. Repeat for your partner's object.

Think About This

1. What kind of clues are the most helpful in identifying an object?

2. How would your clues change if you were describing a substance, such as iron or water, rather than an object?

3. 🔑 **Key Concept** How do you think you use similar clues in your daily life?

Physical Properties

REVIEW VOCABULARY · · · ·

property
a characteristic used to describe something

· · · · · · · · · · · · · · · · ·

As you read in Lesson 1, the arrangement of atoms determines whether matter is a substance or a mixture. The arrangement of atoms also determines the **properties** of different types of matter. Each element and compound has a unique set of properties. When substances mix together and form mixtures, the properties of the substances that make up the mixture are still present.

You can observe some properties of matter, and other properties can be measured. For example, you can see that gold is shiny, and you can find the mass of a sample of iron. Think about how you might describe the different substances and mixtures in the photo on the previous page. Could you describe some of the matter in the photo as a solid or a liquid? Why do the water and the rocks leave the pan before the gold does? Could you describe the mass of the various items in the photo? Each of these questions asks about the physical properties of matter. *A* **physical property** *is a characteristic of matter that you can observe or measure without changing the identity of the matter.* There are many types of physical properties, and you will read about some of them in this lesson.

States of Matter

How do aluminum, water, and air differ? Recall that aluminum is an element, water is a compound, and air is a mixture. How else do these three types of matter differ? At room temperature, aluminum is a solid, water is a liquid, and air is a gas. Solids, liquids, and gases are called states of matter. The state of matter is a physical property of matter. Substances and mixtures can be solids, liquids, or gases. For example, water in the ocean is a liquid, but water in an iceberg is a solid. In addition, water vapor in the air above the ocean is a gas.

Did you know that the particles, or atoms and groups of atoms, that make up all matter are constantly moving and are attracted to each other? Look at your pencil. It is made up of trillions of moving particles. Every solid, liquid, and gas around you is made up of moving particles that attract one another. What makes some matter a solid and other matter a liquid or a gas? It depends on how close the particles in the matter are to one another and how fast they move, as shown in **Figure 8.**

✓ **Reading Check** How do solids, liquids, and gases differ?

Figure 8 The three common states of matter on Earth are solid, liquid, and gas.

((○)) **Concepts in Motion** Animation

Solids, Liquids, and Gases

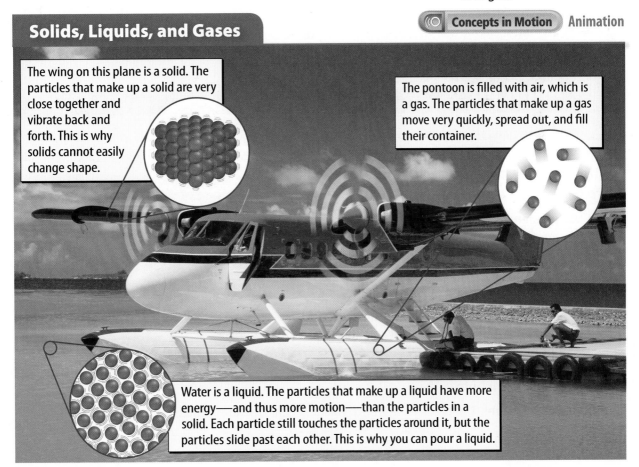

The wing on this plane is a solid. The particles that make up a solid are very close together and vibrate back and forth. This is why solids cannot easily change shape.

The pontoon is filled with air, which is a gas. The particles that make up a gas move very quickly, spread out, and fill their container.

Water is a liquid. The particles that make up a liquid have more energy—and thus more motion—than the particles in a solid. Each particle still touches the particles around it, but the particles slide past each other. This is why you can pour a liquid.

✓ **Visual Check** Which state of matter flows, keeps the same volume, and takes the shape of its container?

Size-Dependent Properties

State is only one of many physical properties that you can use to describe matter. Some physical properties, such as mass and volume, depend on the size or amount of matter. Measurements of these properties vary depending on how much matter is in a sample.

Mass Imagine holding a small dumbbell in one hand and a larger one in your other hand. What do you notice? The larger dumbbell seems heavier. The larger dumbbell has more mass than the smaller one. **Mass** *is the amount of matter in an object.* Both small dumbbells shown in **Figure 9** have the same mass because they both contain the same amount of matter. Mass is a size-dependent property of a given substance because its value depends on the size of a sample.

Mass sometimes is confused with weight, but they are not the same. Mass is an amount of matter in something. Weight is the pull of gravity on that matter. Weight changes with location, but mass does not. Suppose one of the dumbbells in the figure was on the Moon. The dumbbell would have the same mass on the Moon that it has on Earth. However, the Moon's gravity is much less than Earth's gravity, so the weight of the dumbbell would be less on the Moon.

Figure 9 The larger dumbbells have greater mass than the smaller dumbbells because they contain more matter.

Inquiry MiniLab

20 minutes

Can the weight of an object change?

When people go on a diet, both their mass and weight might change. Can the weight of an object change without changing its mass? Let's find out.

1. Read and complete a lab safety form.
2. Use a **balance** to find the mass of five **metal washers.** Record the mass in grams in your Science Journal.
3. Hang the washers from the hook on a **spring scale.** Record the weight in newtons.
4. Lower just the washers into a **500-mL beaker** containing approximately 300 mL water. Record the weight in newtons.

Analyze and Conclude

1. **Draw Conclusions** Did the weight of the washers change during the experiment? How do you know?

2. **Predict** In what other ways might you change the weight of the washers?

3. 🔑 **Key Concept** What factors affect the weight of an object, but not its mass?

Volume Another physical property that depends on the size or the amount of a substance is volume. A unit often used to measure volume is the milliliter (mL). Volume is the amount of space something takes up. Suppose a full bottle of water contains 400 mL of water. If you pour exactly half of the water out, the bottle contains half of the original volume, or 200 mL, of water.

 Reading Check What is a common unit for volume?

Size-Independent Properties

Unlike mass, weight, and volume, some physical properties of a substance do not depend on the amount of matter present. These properties are the same for both small samples and large samples. They are called size-independent properties. Examples of size-independent properties are melting point, boiling point, density, electrical conductivity, and solubility.

Melting Point and Boiling Point The temperature at which a substance changes from a solid to a liquid is its melting point. The temperature at which a substance changes from a liquid to a gas is its boiling point. Different substances have different boiling points and melting points. The boiling point for water is 100°C at sea level. Notice in **Figure 10** that this temperature does not depend on how much water is in the container.

Density Imagine holding a bowling ball in one hand and a foam ball of the same size in the other. The bowling ball seems heavier because the density of the material that makes up the bowling ball is greater than the density of foam. **Density** *is the mass per unit volume of a substance.* Like melting point and boiling point, density is a size-independent property.

Math Skills

Use Ratios

When you compare two numbers by division, you are using a ratio. Density can be written as a ratio of mass and volume. What is the density of a substance if a 5-mL sample has a mass of 25 g?

1. Set up a ratio.

$$\frac{mass}{volume} = \frac{25\text{ g}}{5\text{ mL}}$$

2. Divide the numerator by the denominator to get the mass (in g) of 1 mL.

$$\frac{25\text{ g}}{5\text{ mL}} = \frac{5\text{ g}}{1\text{ mL}}$$

3. The density is 5 g/mL.

Practice

A sample of wood has a mass of 12 g and a volume of 16 mL. What is the density of the wood?

 Review

- **Math Practice**
- **Personal Tutor**

WORD ORIGIN · · · · · · · · · · ·

> **density**
> from Latin *densus*, means "compact"; and Greek *dasys*, means "thick"

Figure 10 The boiling point of water is 100°C at sea level. The boiling point does not change for different volumes of water.

 **Review** **Personal Tutor**

Conductivity Another property that is independent of the sample size is conductivity. Electrical conductivity is the ability of matter to conduct, or carry along, an electric current. Copper often is used for electrical wiring because it has high electrical conductivity. Thermal conductivity is the ability of a material to conduct thermal energy. Metals tend to have high electrical and thermal conductivity. Stainless steel, for example, often is used to make cooking pots because of its high thermal conductivity. However, the handles on the pan probably are made out of wood, plastic, or some other substance that has low thermal conductivity.

Reading Check What are two types of conductivity?

Solubility Have you ever made lemonade by stirring a powdered drink mix into water? As you stir, the powder mixes evenly in the water. In other words, the powder dissolves in the water.

What do you think would happen if you tried to dissolve sand in water? No matter how much you stir, the sand does not dissolve. **Solubility** *is the ability of one substance to dissolve in another.* The powdered drink mix is soluble in water, but sand is not. **Table 1** explains how physical properties such as conductivity and solubility can be used to identify objects and separate mixtures.

Key Concept Check What are five different physical properties of matter?

Table 1 This table contains the descriptions of several physical properties. It also shows examples of how physical properties can be used to separate mixtures.

Visual Check How might you separate a mixture of iron filings and salt?

Concepts in Motion Interactive Table

Table 1 Physical Properties of Matter			
	Property		
	Mass	**Conductivity**	**Volume**
Description of property	The amount of matter in an object	The ability of matter to conduct, or carry along, electricity or heat	The amount of space something occupies
Size-dependent or size independent	Size-dependent	Size-independent	Size-dependent
How the property is used to separate a mixture (example)	Mass typically is not used to separate a mixture.	Conductivity typically is not used to separate a mixture.	Volume could be used to separate mixtures whose parts can be separated by filtration.

Separating Mixtures

In Lesson 1, you read about different types of mixtures. Recall that the substances that make up mixtures are not held together by chemical **bonds.** When substances form a mixture, the properties of the individual substances do not change. One way that a mixture and a compound differ is that the parts of a mixture often can be separated by physical properties. For example, when salt and water form a solution, the salt and the water do not lose any of their individual properties. Therefore, you can separate the salt from the water by using differences in their physical properties. Water has a lower boiling point than salt. If you boil salt water, the water will boil away, and the salt will be left behind. Other physical properties that can be used to separate different mixtures are described in **Table 1.**

Physical properties cannot be used to separate a compound into the elements it contains. The atoms that make up a compound are bonded together and cannot be separated by physical means. For example, you cannot separate the hydrogen atoms from the oxygen atoms in water by boiling water.

 Key Concept Check How are physical properties used to separate mixtures?

SCIENCE USE v. COMMON USE

bond
Science Use a force between atoms or groups of atoms

Common Use a monetary certificate issued by a government or a business that earns interest

Property

Boiling/Melting Points	State of matter	Density	Solubility	Magnetism
The temperature at which a material changes state	Whether something is a solid, a liquid, or a gas	The amount of mass per unit of volume	The ability of one substance to dissolve in another	Attractive force for some metals, especially iron
Size-independent	Size-independent	Size-independent	Size-independent	Size-independent
Each part of a mixture will boil or melt at a different temperature.	A liquid can be poured off a solid.	Objects with greater density sink in objects with less density.	Dissolve a soluble material to separate it from a material with less solubility.	Attract iron from a mixture of materials.

Lesson 2 Review

Visual Summary

A physical property is a characteristic of matter that can be observed or measured without changing the identity of the matter.

Examples of physical properties include mass, density, volume, melting point, boiling point, state of matter, and solubility.

Many physical properties can be used to separate the components of a mixture.

FOLDABLES

Use your lesson Foldable to review the lesson. Save your Foldable for the project at the end of the chapter.

What do you think NOW?

You first read the statements below at the beginning of the chapter.

3. The weight of a material never changes, regardless of where it is.

4. Boiling is one method used to separate parts of a mixture.

Did you change your mind about whether you agree or disagree with the statements? Rewrite any false statements to make them true.

Use Vocabulary

1 **Distinguish** between mass and weight.

2 **Use the term** *solubility* in a sentence.

3 An object's _____ is the amount of mass per a certain unit of volume.

Understand Key Concepts

4 **Explain** how to separate a mixture of sand and pebbles.

5 Which physical property is NOT commonly used to separate mixtures?
   A. magnetism   **C.** density
   B. conductivity   **D.** solubility

6 **Analyze** Name two size-dependent properties and two size-independent properties of an iron nail.

Interpret Graphics

7 **Sequence** Draw a graphic organizer like the one below to show the steps in separating a mixture of sand, iron filings, and salt.

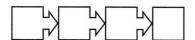

Critical Thinking

8 **Examine** the diagram below.

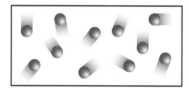

How can you identify the state of matter represented by the diagram?

Math Skills

Review — Math Practice

9 A piece of copper has a volume of 100.0 cm³. If the mass of the copper is 890 g, what is the density of copper?

How can following a procedure help you solve a crime?

Materials

Plastic sealable bag

triple-beam balance

50-mL graduated cylinder

paper towels

Also needed:

Crime Scene Objects

Imagine that you are investigating a crime scene. You find several pieces of metal and broken pieces of plastic that look as if they came from a car's tail light. You also have similar objects collected from the suspect. How can you figure out if they are parts of the same objects?

Learn It

To be sure you do the same tests on each object, it is helpful to **follow a procedure.** A procedure tells you how to use the materials and what steps to take.

Try It

1 Read and complete a lab safety form.

2 Copy the table below into your Science Journal.

3 Use the balance to find the mass of an object from the crime scene. Record the mass in your table.

4 Place about 25 mL of water in a graduated cylinder. Read and record the exact volume. Call this volume V_1.

5 Carefully tilt the cylinder, and allow one of the objects to slide into the water. Read and record the volume. Call this volume V_2.

6 Repeat steps 3–5 for each of the other objects.

Apply It

7 Complete the table by calculating the volume and the density of each object.

8 What conclusions can you draw about the objects collected from the crime scene and those collected from the suspect?

9 **Key Concept** How could you use this procedure to help identify and compare various objects?

Object	Mass (M) (g)	V_1 (mL)	V_2 (mL)	Volume of Object (V) ($V_2 - V_1$) (mL)	Density of Object M/V (g/mL)
1					
2					
3					
4					
5					
6					

Reading Guide

Key Concepts
ESSENTIAL QUESTIONS

- How can a change in energy affect the state of matter?
- What happens when something dissolves?
- What is meant by conservation of mass?

Vocabulary
physical change p. 249

g Multilingual eGlossary

Physical Changes

Inquiry Change by Chipping?

This artist is changing a piece of wood into an instrument that will make beautiful music. He planned and chipped, measured and shaped. Chips of wood flew, and rough edges became smooth. Although the wood changed shape, it remained wood. Its identity did not change, just its form.

Where did it go?

When you dissolve sugar in water, where does the sugar go? One way to find out is to measure the mass of the water and the sugar before and after mixing.

1. Read and complete a lab safety form.

2. Add **sugar** to a **small paper cup** until the cup is approximately half full. Bend the cup's opening, and pour the sugar into a **balloon.**

3. With the balloon hanging over the side, stretch the neck of the balloon over a **flask** half full of **water.**

4. Use a **balance** to find the mass of the flask-and-balloon assembly. Record the mass in your Science Journal.

5. Lift the end of the balloon, and empty the sugar into the flask. Swirl until the sugar dissolves. Measure and record the mass of the flask-and-balloon assembly again.

Think About This

1. Is the sugar still present after it dissolves? How do you know?

2. 🔑 **Key Concept** Based on your observations, what do you think happens to the mass of objects when they dissolve? Explain.

Physical Changes

How would you describe water? If you think about water in a stream, you might say that it is a cool liquid. If you think about water as ice, you might describe it as a cold solid. How would you describe the change from ice to water? As ice melts, some of its properties change, such as the state of matter, the shape, and the temperature, but it is still water. In Lesson 2, you read that substances and mixtures can be solids, liquids, or gases. In addition, substances and mixtures can change from one state to another. *A* **physical change** *is a change in size, shape, form, or state of matter in which the matter's identity stays the same.* During a physical change, the matter does not become something different even though physical properties change.

Change in Shape and Size

Think about changes in the shapes and the sizes of substances and mixtures you experience each day. When you chew food, you are breaking it into smaller pieces. This change in size helps make food easier to digest. When you pour juice from a bottle into a glass, you are changing the shape of the juice. If you fold clothes to fit them into a drawer, you are changing their shapes. Changes in shape and size are physical changes. The identity of the matter has not changed.

WORD ORIGIN

physical
from Greek *physika*, means "natural things"

change
from Latin *cambire*, means "to exchange"

FOLDABLES

Make a vertical two-tab book. Label the tabs as illustrated. Record specific examples illustrating how adding or releasing thermal energy results in physical change.

Increasing Thermal Energy

Decreasing Thermal Energy

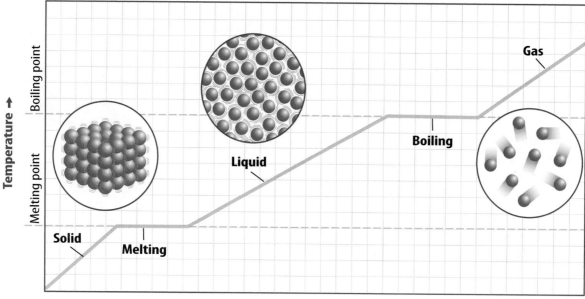

Adding thermal energy ➡

▲ Figure 11 🔑 As thermal energy is added to a material, temperature increases when the state of the material is not changing. Temperature stays the same during a change of state.

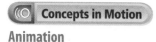

Animation

Figure 12 Solid iodine undergoes sublimation. It changes from a solid to a gas without becoming a liquid. ▼

Change in State of Matter

Why does ice melt in your hand? Or, why does water turn to ice in the freezer? Matter, such as water, can change state. Recall from Lesson 2 how the particles in a solid, a liquid, and a gas behave. To change the state of matter, the movement of the particles has to change. In order to change the movement of particles, thermal energy must be either added or removed.

Adding Thermal Energy When thermal energy is added to a solid, the particles in the solid move faster and faster, and the temperature increases. As the particles move faster, they are more likely to overcome the attractive forces that hold them tightly together. When the particles are moving too fast for attractive forces to hold them tightly together, the solid reaches its melting point. The melting point is the temperature at which a solid changes to a liquid.

After all the solid has melted, adding more thermal energy causes the particles to move even faster. The temperature of the liquid increases. When the particles are moving so fast that attractive forces cannot hold them close together, the liquid is at its boiling point. The boiling point is the temperature at which a liquid changes into a gas and the particles spread out. **Figure 11** shows how temperature and change of state relate to each other when thermal energy is added to a material.

Some solids change directly to a gas without first becoming a liquid. This process is called sublimation. An example of sublimation is shown in **Figure 12.** You saw another example of sublimation in **Figure 5** in Lesson 1.

Removing Thermal Energy When thermal energy is removed from a gas, such as water vapor, particles in the gas move more slowly and the temperature decreases. Condensation occurs when the particles are moving slowly enough for attractive forces to pull the particles close together. Recall that condensation is the process that occurs when a gas becomes a liquid.

After the gas has completely changed to a liquid, removing more thermal energy from the liquid causes particles to move even more slowly. As the motion between the particles slows, the temperature decreases. Freezing occurs when the particles are moving so slowly that attractive forces between the particles hold them tightly together. Now the particles only can vibrate in place. Recall that freezing is the process that occurs when a liquid becomes a solid.

Freezing and melting are reverse processes, and they occur at the same temperature. The same is true of boiling and condensation. Another change of state is deposition. Deposition is the change from a gas directly to a solid, as shown in **Figure 13.** It is the process that is the opposite of sublimation.

 Key Concept Check How can removing thermal energy affect the state of matter?

 **Inquiry MiniLab**   **30 minutes**

Can you make ice without a freezer?

What happens when you keep removing energy from a substance?

1. Read and complete a lab safety form.

2. Draw the data table below in your Science Journal. Half fill a **large test tube** with **distilled water.** Use a **thermometer** to measure the temperature of the water, then record it.

3. Place the test tube into a **large foam cup** containing **ice** and **salt.** Use a **stirring rod** to slowly stir the water in the tube.

4. Record the temperature of the water every minute until the water freezes. Continue to record the temperature each minute until it drops to several degrees below 0°C.

Time (min)	0	1	2	3	4	5	6	7	8
Temperature (C)									

Analyze and Conclude

1. **Organize Data** Graph the data in your table. Label time on the *x*-axis and temperature on the *y*-axis.

2. **Interpret Data** According to your data, what is the freezing point of water?

3. **Key Concept** What caused the water to freeze?

Figure 13 When enough thermal energy is removed, one of several processes occurs.

Freezing

Condensation

Deposition

▲ **Figure 14** Salt dissolves when it is added to the water in this aquarium.

Dissolving

Have you ever owned a saltwater aquarium, such as the one shown in **Figure 14?** If you have, you probably had to add certain salts to the water before you added the fish. Can you see the salt in the water? As you added the salt to the water, it gradually disappeared. It was still there, but it dissolved, or mixed evenly, in the water. Because the identities of the substances—water and salt—are not changed, dissolving is a physical change.

Like many physical changes, dissolving is usually easy to reverse. If you boil the salt water, the liquid water will change to water vapor, leaving the salt behind. You once again can see the salt because the particles that make up the substances do not change identity during a physical change.

Key Concept Check What happens when something dissolves?

Conservation of Mass

During a physical change, the physical properties of matter change. The particles in matter that are present before a physical change are the same as those present after the physical change. Because the particles are the same both before and after a physical change, the total mass before and after the change is also the same, as shown in **Figure 15.** This is known as the conservation of mass. You will read in Lesson 4 that mass also is conserved during another type of change—a chemical change.

Key Concept Check What is meant by conservation of mass?

Figure 15 Mass is conserved during a physical change. ▼

Conservation of Mass

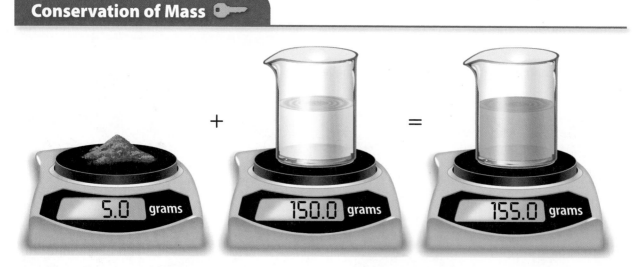

Visual Check If a sample of water has a mass of 200 g and the final solution has a mass of 230 g, how much solute dissolved in the water?

Lesson 3 Review

Visual Summary

During a physical change, matter can change form, shape, size, or state, but the identity of the matter does not change.

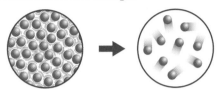

 Matter either changes temperature or changes state when enough thermal energy is added or removed.

 Mass is conserved during physical changes, which means that mass is the same before and after the changes occur.

FOLDABLES®

Use your lesson Foldable to review the lesson. Save your Foldable for the project at the end of the chapter.

What do you think NOW?

You first read the statements below at the beginning of the chapter.

5. Heating a material decreases the energy of its particles.

6. When you stir sugar into water, the sugar and water evenly mix.

Did you change your mind about whether you agree or disagree with the statements? Rewrite any false statements to make them true.

Use Vocabulary

1 **Use the term** *physical change* in a sentence.

Understand Key Concepts

2 **Describe** how a change in energy can change ice into liquid water.

3 Which never changes during a physical change?
   A. state of matter   C. total mass
   B. temperature   D. volume

4 **Relate** What happens when something dissolves?

Interpret Graphics

5 **Examine** the graph below of temperature over time as a substance changes from solid to liquid to gas. Explain why the graph has horizontal lines.

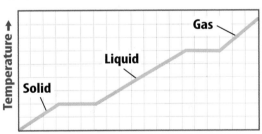

6 **Take Notes** Copy the graphic organizer below. For each heading, summarize the main idea described in the lesson.

Heading	Main Idea
Physical Changes	
Change in State of Matter	
Conservation of Mass	

Critical Thinking

7 **Design** a demonstration that shows that temperature remains unchanged during a change of state.

How can known substances help you identify unknown substances?

Materials

plastic spoons

magnifying lens

stirring rod

Also needed: known substances (baking soda, ascorbic acid, sugar, cornstarch) test tubes, test tube rack, watch glass, dropper bottles containing water, iodine, vinegar, and red cabbage indicator

Safety

While investigating a crime scene, you find several packets of white powder. Are they illegal drugs or just harmless packets of candy? Here's one way to find out.

Learn It

A **control** is something that stays the same. If you determine how a known substance reacts with other substances, you can use it as a control. Unknown substances are **variables.** They might or might not react in the same way.

Try It

1 Read and complete a lab safety form.

2 Copy the data table below into your Science Journal.

3 Use a magnifying lens to observe the appearance of each known substance.

4 Test small samples of each known substance for their reaction with a drop or two of water, vinegar, and iodine solution.

5 Feel the texture of each substance.

6 Mix each substance with water, and add the red cabbage indicator.

7 After you complete your observations, ask your teacher for a mystery powder. Repeat steps 3–6 using the mystery powder. Use the data you collect to identify the powder.

Apply It

8 What test suggests that a substance might be cornstarch?

9 Why should you test the reactions of the substances with many different things?

10 🔑 **Key Concept** How did you use the properties of the controls to identify your variable?

Substance	Appearance	Texture	Reaction to Water	Reaction to Iodine	Reaction to Vinegar	Red Cabbage Indicator
Baking soda						
Sugar						
Ascorbic acid						
Cornstarch						
Mystery powder						

Lesson 4

Reading Guide

Key Concepts 🔑
ESSENTIAL QUESTIONS

- What is a chemical property?
- What are some signs of chemical change?
- Why are chemical equations useful?
- What are some factors that affect the rate of chemical reactions?

Vocabulary

chemical property p. 256

chemical change p. 257

concentration p. 260

 Multilingual eGlossary

 Video

- BrainPOP®
- What's Science Got to do With It?

Chemical Properties and Changes

Inquiry A Burning Issue?

As this car burns, some materials change to ashes and gases. The metal might change form or state if the fire is hot enough, but it probably won't burn. Why do fabric, leather, and paint burn? Why do many metals not burn? The properties of matter determine how matter behaves when it undergoes a change.

What can colors tell you?

You mix red and blue paint to get purple paint. Iron changes color when it rusts. Are color changes physical changes?

1. Read and complete a lab safety form.

2. Divide a **paper towel** into thirds. Label one section *RCJ*, the second section *A*, and the third section *B*.

3. Dip one end of three **cotton swabs** into **red cabbage juice** (RCJ). Observe the color, and set the swabs on the paper towel, one in each of the three sections.

4. Add one drop of **substance A** to the swab in the *A* section. Observe any changes, and record observations in your Science Journal.

5. Repeat step 4 with **substance B** and the swab in the *B* section.

6. Observe **substances C** and **D** in their **test tubes**. Then pour C into D. Rock the tube gently to mix. Record your observations.

Think About This

1. What happened to the color of the red cabbage juice when substances A and B were added?

2.  **Key Concept** Which of the changes you observed do you think was a physical change? Explain your reasoning.

Chemical Properties

Recall that a physical property is a characteristic of matter that you can observe or measure without changing the identity of the matter. However, matter has other properties that can be observed only when the matter changes from one substance to another. *A* **chemical property** *is a characteristic of matter that can be observed as it changes to a different type of matter.* For example, what are some chemical properties of a piece of paper? Can you tell by just looking at it that it will burn easily? The only way to know that paper burns is to bring a flame near the paper and watch it burn. When paper burns, it changes into different types of matter. The ability of a substance to burn is a chemical property. The ability to rust is another chemical property.

Comparing Properties

You now have read about physical properties and chemical properties. All matter can be described using both types of properties. For example, a wood log is solid, rounded, heavy, and rough. These are physical properties that you can observe with your senses. The log also has mass, volume, and density, which are physical properties that can be measured. The ability of wood to burn is a chemical property. This property is obvious only when you burn the wood. It also will rot, another chemical property you can observe when the log decomposes, becoming other substances. When you describe matter, you consider both its physical and its chemical properties.

Key Concept Check What are some chemical properties of matter?

Chemical Changes

Recall that during a physical change, the identity of matter does not change. However, *a* **chemical change** *is a change in matter in which the substances that make up the matter change into other substances with new physical and chemical properties.* For example, when iron undergoes a chemical change with oxygen, rust forms. The substances that undergo a change no longer have the same properties because they no longer have the same identity.

 Reading Check What is the difference between a physical change and a chemical change?

Signs of Chemical Change

How do you know when a chemical change occurs? What signs show you that new types of matter form? As shown in **Figure 16,** signs of chemical changes include the formation of bubbles or a change in odor, color, or energy.

It is important to remember that these signs do not always mean a chemical change occurred. Think about what happens when you heat water on a stove. Bubbles form as the water boils. In this case, bubbles show that the water is changing state, which is a physical change. The evidence of chemical change shown in **Figure 16** means that a chemical change might have occurred. However, the only proof of chemical change is the formation of a new substance.

Key Concept Check What are signs of a chemical change?

FOLDABLES

Use a sheet of paper to make a chart with four columns. Use the chart throughout this lesson to explain how the identity of matter changes during a chemical change.

Action/ Matter	Signs of Chemical Change	Explain the Chemical Reaction	What affects the reaction rate?

WORD ORIGIN

chemical
from Greek *chemeia*, means "cast together"

Some Signs of Chemical Change

Figure 16 Sometimes you can observe clues that a chemical change has occurred.

Bubbles

Energy change

Odor change

Color change

 Visual Check What signs show that a chemical change takes place when fireworks explode?

Can you spot the clues for chemical change?

What are some clues that let you know a chemical change might have taken place?

1. Read and complete a lab safety form.

2. Add about 25 mL of room-temperature water to a **self-sealing plastic bag.** Add two **dropperfuls** of **red cabbage juice.**

3. Add one **measuring scoop** of **calcium chloride** to the bag. Seal the bag. Tilt the bag to mix the contents until the solid disappears. Feel the bottom of the bag. Record your observations in your Science Journal.

4. Open the bag, and add one measuring scoop of **baking soda.** Quickly press the air from the bag and reseal it. Tilt the bag to mix the contents. Observe for several minutes. Record your observations.

Analyze and Conclude

1. **Observe** What changes did you observe?

2. **Infer** Which of the changes suggested that a new substance formed? Explain.

3. **Key Concept** Are changes in energy always a sign of a chemical change? Explain.

Explaining Chemical Reactions

You might wonder why chemical changes produce new substances. Recall that particles in matter are in constant motion. As particles move, they collide with each other. If the particles collide with enough force, the bonded atoms that make up the particles can break apart. These atoms then rearrange and bond with other atoms. When atoms bond together in new combinations, new substances form. This process is called a reaction. Chemical changes often are called chemical reactions.

Reading Check What does it mean to say that atoms rearrange during a chemical change?

Using Chemical Formulas

A useful way to understand what happens during a chemical reaction is to write a chemical equation. A chemical equation shows the chemical formula of each substance in the reaction. The formulas to the left of the arrow represent the reactants. Reactants are the substances present before the reaction takes place. The formulas to the right of the arrow represent the products. Products are the new substances present after the reaction. The arrow indicates that a reaction has taken place.

Key Concept Check Why are chemical equations useful?

Figure 17 Chemical formulas and other symbols are parts of a chemical equation.

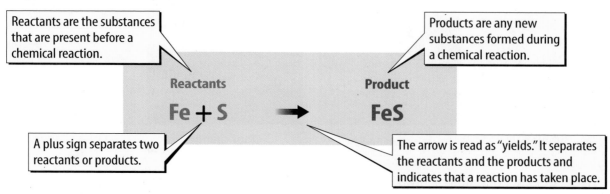

Reactants are the substances that are present before a chemical reaction.

Products are any new substances formed during a chemical reaction.

Reactants

$Fe + S$

Product

FeS

A plus sign separates two reactants or products.

The arrow is read as "yields." It separates the reactants and the products and indicates that a reaction has taken place.

Balancing Chemical Equations

Look at the equation in **Figure 17**. Notice that there is one iron (Fe) atom on the reactants side and one iron atom on the product side. This is also true for the sulfur (S) atoms. Recall that during both physical and chemical changes, mass is conserved. This means that the total mass before and after a change must be equal. Therefore, in a chemical equation, the number of atoms of each element before a reaction must equal the number of atoms of each element after the reaction. This is called a balanced chemical equation, and it illustrates the conservation of mass. **Figure 18** explains how to write and balance a chemical equation.

When balancing an equation, you cannot change the chemical formula of any reactants or products. Changing a formula changes the identity of the substance. Instead, you can place coefficients, or multipliers, in front of formulas. Coefficients change the amount of the reactants and products present. For example, an H_2O molecule has two H atoms and one O atom. Placing the coefficient 2 before H_2O ($2H_2O$) means that you double the number of H atoms and O atoms present:

$$2 \times 2 \text{ H atoms} = 4 \text{ H atoms}$$
$$2 \times 1 \text{ O atom} = 2 \text{ O atoms}$$

Note that $2H_2O$ is still water. However, it describes two water particles instead of one.

Figure 18 Equations must be balanced because mass is conserved during a chemical reaction.

Balancing Chemical Equations

 Review   Personal Tutor

Balancing Chemical Equations Example
When methane (CH_4)—a gas burned in furnaces—reacts with oxygen (O_2) in the air, the reaction produces carbon dioxide (CO_2) and water (H_2O). Write and balance a chemical equation for this reaction.

1 **Write the equation, and check to see if it is balanced.**

a. Write the chemical formulas with the reactants on the left side of the arrow and the products on the right side.	**a.** $CH_4 + O_2 \rightarrow CO_2 + H_2O$ **not balanced**

b. Count the atoms of each element in the reactants and in the products.
 ■ Note which elements have a balanced number of atoms on each side of the equation.
 ■ If all elements are balanced, the overall equation is balanced. If not, go to step 2.
	b. reactants $\rightarrow$ products
	     C=1 C=1 **balanced**
	     H=4 H=2 **not balanced**
	     O=2 O=3 **not balanced**

2 **Add coefficients to the chemical formulas to balance the equation.**

a. Pick an element in the equation whose atoms are not balanced, such as hydrogen. Write a coefficient in front of a reactant or a product that will balance the atoms of the chosen element in the equation.
 	
 b. Recount the atoms of each element in the reactants and the products, and note which are balanced on each side of the equation.
 	
 c. Repeat steps 2a and 2b until all atoms of each element in the reactants equal those in the products.
	a. $CH_4 + O_2 \rightarrow CO_2 + 2H_2O$ **not balanced**
	 b. C=1 C=1 **balanced**
	      H=4 H=4 **balanced**
	      O=2 O=4 **not balanced**
	 c. $CH_4 + 2O_2 \rightarrow CO_2 + 2H_2O$ **balanced**
	        C=1 C=1 **balanced**
	        H=4 H=4 **balanced**
	        O=4 O=4 **balanced**

3 **Write the balanced equation that includes the coefficients:** $CH_4 + 2O_2 \rightarrow CO_2 + 2H_2O$

Factors that Affect the Rate of Chemical Reactions 🔑

Figure 19 The rate of most chemical reactions increases with an increase in temperature, concentration, or surface area.

1 **Temperature**

Chemical reactions that occur during cooking happen at a faster rate when temperature increases.

2 **Concentration**

Acid rain contains a higher concentration of acid than normal rain does. As a result, a statue exposed to acid rain is damaged more quickly than a statue exposed to normal rain.

3 **Surface Area**

When an antacid tablet is broken into pieces, the pieces have more total surface area than the whole tablet does. The pieces react more rapidly with water because more of the broken tablet is in contact with the water.

The Rate of Chemical Reactions

Recall that the particles that make up matter are constantly moving and colliding with one another. Different factors can make these particles move faster and collide harder and more frequently. These factors increase the rate of a chemical reaction, as shown in **Figure 19.**

1 A higher **temperature** usually increases the rate of reaction. When the temperature is higher, the particles move faster. Therefore, the particles collide with greater force and more frequently.

2 **Concentration** *is the amount of substance in a certain volume.* A reaction occurs faster if the concentration of at least one reactant increases. When concentration increases, there are more particles available to bump into each other and react.

3 **Surface area** also affects reaction rate if at least one reactant is a solid. If you drop a whole effervescent antacid tablet into water, the tablet reacts with the water. However, if you break the tablet into several pieces and then add them to the water, the reaction occurs more quickly. Smaller pieces have more total surface area, so more space is available for reactants to collide.

 Key Concept Check List three factors that affect the rate of a chemical reaction.

Chemistry

To understand chemistry, you need to understand matter. You need to know how the arrangement of atoms results in different types of matter. You also need to be able to distinguish physical properties from chemical properties and describe ways these properties can change. In later chemistry chapters and courses, you will examine each of these topics closely to gain a better understanding of matter.

Lesson 4 Review

Visual Summary

A chemical property is observed only as a material undergoes chemical change and changes identity.

Signs of possible chemical change include bubbles, energy change, and change in odor or color.

Chemical equations show the reactants and products of a chemical reaction and that mass is conserved.

Reactants   Product

Fe + S ➔ FeS

FOLDABLES

Use your lesson Foldable to review the lesson. Save your Foldable for the project at the end of the chapter.

What do you think NOW?

You first read the statements below at the beginning of the chapter.

7. When wood burns, new materials form.

8. Temperature can affect the rate at which chemical changes occur.

Did you change your mind about whether you agree or disagree with the statements? Rewrite any false statements to make them true.

Use Vocabulary

1 The amount of substance in a certain volume is its _____.

2 **Use the term** *chemical change* in a complete sentence.

Understand Key Concepts 🔑

3 **List** some signs of chemical change.

4 Which property of matter changes during a chemical change but does NOT change during a physical change?
- **A.** energy
- **B.** identity
- **C.** mass
- **D.** volume

5 **State** why chemical equations are useful.

6 **Analyze** What affects the rate at which acid rain reacts with a statue?

Interpret Graphics

7 **Examine** Explain how the diagram below shows conservation of mass.

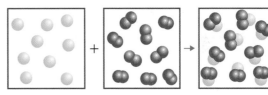

8 **Compare and Contrast** Copy and fill in the graphic organizer to compare and contrast physical and chemical changes.

Physical and Chemical Changes	
Alike	
Different	

Critical Thinking

9 **Compile** a list of three physical changes and three chemical changes you have observed recently.

10 **Recommend** How could you increase the rate at which the chemical reaction between vinegar and baking soda occurs?

Materials

triple-beam balance

50-mL graduated cylinder

magnifying lens

bar magnet

Also needed:
crime scene evidence, unknown substances, dropper bottles containing water, iodine, cornstarch, and red cabbage indicator, test tubes, test tube rack, stirring rod

Safety

Design an Experiment to Solve a Crime

Recall how you can use properties to identify and compare substances. You now will apply those ideas to solving a crime. You will be given evidence collected from the crime scene and from the suspect's house. As the investigator, decide whether evidence from the crime scene matches evidence from the suspect. What tests will you use? What does the evidence tell you?

Question

Determine which factors about the evidence you would like to investigate further. Consider how you can describe and compare the properties of each piece of evidence. Evaluate the properties you will observe and measure, and decide whether it would be an advantage to classify them as physical properties or chemical properties. Will the changes that the evidence will undergo be helpful to you? Think about controls, variables, and the equipment you have available. Is there any way to match samples exactly?

Procedure

1 Read and complete a lab safety form.

2 In your Science Journal, write the procedures you will use to answer your question. Include the materials and steps you will use to test each piece of evidence. By the appropriate step in the procedure, list any safety procedures you should observe while performing the investigation. Organize your steps by putting them in a graphic organizer, such as the one below. Have your teacher approve your procedures.

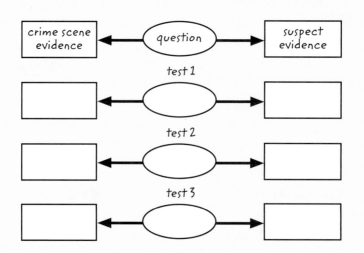

3 Begin by observing and recording your observations on each piece of evidence. What can you learn by comparing physical properties? Are any of the samples made of several parts?

4 Use the available materials to test the evidence. Accurately record all observations and data for each piece of evidence.

5 Add any additional tests you think you need to answer your questions.

Analyze and Conclude

6 Examine the data you have collected. What does the evidence tell you about whether the crime scene and the suspect are related?

7 Write your conclusions in your Science Journal. Be thorough because these are the notes you would use if you had to testify in court about the case.

8 **Analyze** Which data suggest that evidence from the crime scene was or wasn't connected to the suspect?

9 **Draw Conclusions** If you were to testify in court, what conclusions would you be able to state confidently based on your findings?

10 **The Big Idea** How does understanding physical and chemical properties of matter help you to solve problems?

Communicate Your Results

Compare your results with those of other teams. Discuss the kinds of evidence that might be strong enough to convict a suspect.

 **Extension**

Research the difference between individual and class evidence used in forensics. Decide which class of evidence your tests provided.

Lab Tips

☑ Don't overlook simple ideas such as matching the edges of pieces.

☑ Can you separate any of the samples into other parts?

☑ Always get your teacher's approval before trying any new test.

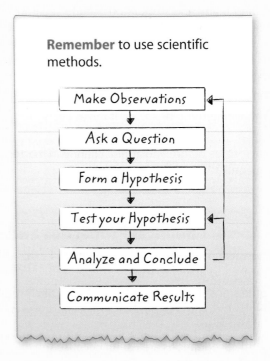

Remember to use scientific methods.

Make Observations

↓

Ask a Question

↓

Form a Hypothesis

↓

Test your Hypothesis

↓

Analyze and Conclude

↓

Communicate Results

Chapter 7 Study Guide

Matter is anything that has mass and takes up space. Its physical properties and its chemical properties can change.

Key Concepts Summary

| | |

Lesson 1: Classifying Matter

- A **substance** is a type of **matter** that always is made of atoms in the same combinations.
- **Atoms** of different elements have different numbers of protons.
- The composition of a substance cannot vary. The composition of a **mixture** can vary.
- Matter can be classified as either a substance or a mixture.

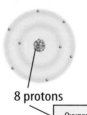

8 protons

Oxygen
8
O
16.00

Lesson 2: Physical Properties

- **Physical properties** of matter include size, shape, texture, and state.
- Physical properties such as **density,** melting point, boiling point, and size can be used to separate mixtures.

Lesson 3: Physical Changes

- A change in energy can change the state of matter.
- When something dissolves, it mixes evenly in a substance.
- The masses before and after a change in matter are equal.

Lesson 4: Chemical Properties and Changes

- **Chemical properties** include ability to burn, acidity, and ability to rust.
- Some signs that might indicate **chemical changes** are the formation of bubbles and a change in odor, color, or energy.
- Chemical equations are useful because they show what happens during a chemical reaction.
- Some factors that affect the rate of chemical reactions are temperature, **concentration,** and surface area.

Vocabulary

matter p. 231
atom p. 231
substance p. 233
element p. 233
compound p. 234
mixture p. 235
heterogeneous mixture p. 235
homogeneous mixture p. 235
dissolve p. 235

physical property p. 240
mass p. 242
density p. 243
solubility p. 244

physical change p. 249

chemical property p. 256
chemical change p. 257
concentration p. 260

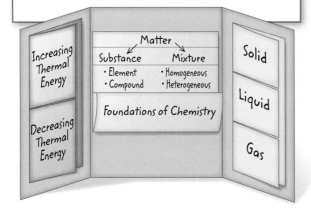
FOLDABLES® Chapter Project

Assemble your lesson Foldables as shown to make a Chapter Project. Use the project to review what you have learned in this chapter. Fasten the Foldable from Lesson 4 on the back of the board.

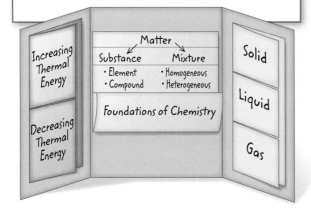

Use Vocabulary

Give two examples of each of the following.

1. element
2. compound
3. homogeneous mixture
4. heterogeneous mixture
5. physical property
6. chemical property
7. physical change
8. chemical change

Link Vocabulary and Key Concepts

Concepts in Motion Interactive Concept Map

Copy this concept map, and then use vocabulary terms from the previous page to complete the concept map.

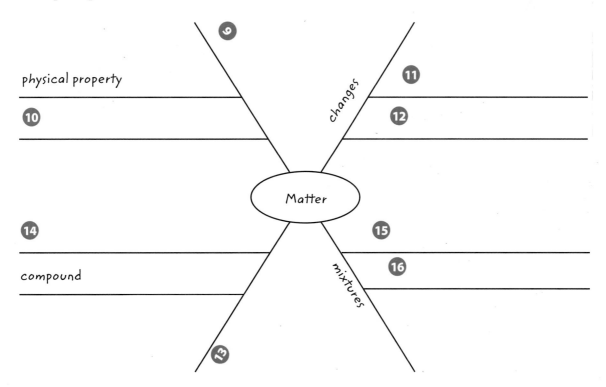

Understand Key Concepts

1 The formula $AgNO_3$ represents a compound made of which atoms?

A. 1 Ag, 1 N, 1 O
B. 1 Ag, 1 N, 3 O
C. 1 Ag, 3 N, 3 O
D. 3 Ag, 3 N, 3 O

2 Which is an example of an element?

A. air
B. water
C. sodium
D. sugar

3 Which property explains why copper often is used in electrical wiring?

A. conductivity
B. density
C. magnetism
D. solubility

4 The table below shows densities for different substances.

Substance	Density (g/cm³)
1	1.58
2	0.32
3	1.52
4	1.62

For which substance would a 4.90-g sample have a volume of 3.10 cm³?

A. substance 1
B. substance 2
C. substance 3
D. substance 4

5 Which would decrease the rate of a chemical reaction?

A. increase in concentration
B. increase in temperature
C. decrease in surface area
D. increase in both surface area and concentration

6 Which physical change is represented by the diagram below?

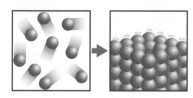

A. condensation
B. deposition
C. evaporation
D. sublimation

7 Which chemical equation is unbalanced?

A. $2KClO_3 \rightarrow 2KCl + 3O_2$
B. $CH_4 + 2O_2 \rightarrow CO_2 + 2H_2O$
C. $Fe_2O_3 + CO \rightarrow 2Fe + 2CO_2$
D. $H_2CO_3 \rightarrow H_2O + CO_2$

8 Which is a size-dependent property?

A. boiling point
B. conductivity
C. density
D. mass

9 Why is the following chemical equation said to be balanced?

$$O_2 + 2PCl_3 \rightarrow 2POCl_3$$

A. There are more reactants than products.
B. There are more products than reactants.
C. The atoms are the same on both sides of the equation.
D. The coefficients are the same on both sides of the equation.

10 The elements sodium (Na) and chlorine (Cl) react and form the compound sodium chloride (NaCl). Which is true about the properties of these substances?

A. Na and Cl have the same properties.
B. NaCl has the properties of Na and Cl.
C. All the substances have the same properties.
D. The properties of NaCl are different from the properties of Na and Cl.

Critical Thinking

11 **Compile** a list of ten materials in your home. Classify each material as an element, a compound, or a mixture.

12 **Evaluate** Would a periodic table based on the number of electrons in an atom be as effective as the one shown in the back of this book? Why or why not?

13 **Develop** a demonstration to show how weight is not the same thing as mass.

14 **Construct** an explanation for how the temperature and energy of a material changes during the physical changes represented by the diagram below.

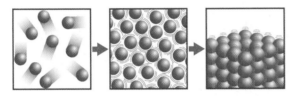

15 **Revise** the definition of physical change given in this chapter so it mentions the type and arrangement of atoms.

16 **Find an example** of a physical change in your home or school. Describe the changes in physical properties that occur during the change. Then explain how you know the change is not a chemical change.

17 **Develop** a list of five chemical reactions you observe each day. For each, describe one way that you could either increase or decrease the rate of the reaction.

Writing in Science

18 **Write** a poem at least five lines long to describe the organization of matter by the arrangement of its atoms. Be sure to include both the names of the different types of matter as well as their meanings.

REVIEW THE BIG IDEA

19 Explain how you are made of matter that undergoes changes. Provide specific examples in your explanation.

20 How does the photo below show an example of a physical change, a chemical change, a physical property, and a chemical property?

Math Skills

Review — Math Practice

Use Ratios

21 A sample of ice at 0°C has a mass of 23 g and a volume of 25 cm³. Why does ice float on water? (The density of water is 1.00 g/cm³.)

22 The table below shows the masses and the volumes for samples of two different elements.

Element	Mass (g)	Volume (cm³)
Gold	386	20
Lead	22.7	2.0

Which element sample in the table has greater density?

Standardized Test Practice

Record your answers on the answer sheet provided by your teacher or on a sheet of paper.

Multiple Choice

1 Which describes how mixtures differ from substances?

   A Mixtures are homogeneous.

   B Mixtures are liquids.

   C Mixtures can be separated physically.

   D Mixtures contain only one kind of atom.

Use the figure below to answer question 2.

A   **B**   **C**   **D**

2 Which image in the figure above is a model for a compound?

   A A

   B B

   C C

   D D

3 Which is a chemical property?

   A the ability to be compressed

   B the ability to be stretched into thin wire

   C the ability to melt at low temperature

   D the ability to react with oxygen

4 You drop a sugar cube into a cup of hot tea. What causes the sugar to disappear in the tea?

   A It breaks into elements.

   B It evaporates.

   C It melts.

   D It mixes evenly.

5 Which is an example of a substance?

   A air

   B lemonade

   C soil

   D water

Use the figure below to answer question 6.

6 The figure above is a model of atoms in a sample at room temperature. Which physical property does this sample have?

   A It can be poured.

   B It can expand to fill its container.

   C It cannot easily change shape.

   D It has a low boiling point.

7 Which observation is a sign of a chemical change?

   A bubbles escaping from a carbonated drink

   B iron filings sticking to a magnet

   C lights flashing from fireworks

   D water turning to ice in a freezer

8 Zinc, a solid metal, reacts with a hydro-chloric acid solution. Which will increase the reaction rate?

 A cutting the zinc into smaller pieces

 B decreasing the concentration of the acid

 C lowering the temperature of the zinc

 D pouring the acid into a larger container

Use the figure below to answer question 9.

9 In the figure above, what will be the mass of the final solution if the solid dissolves in the water?

 A 5 g

 B 145 g

 C 150 g

 D 155 g

10 Which is NOT represented in a chemical equation?

 A chemical formula

 B product

 C conservation of mass

 D reaction rate

Constructed Response

Use the graph below to answer questions 11 and 12.

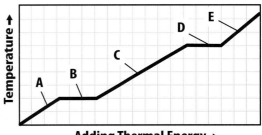

Adding Thermal Energy →

11 Use the graph above to explain why ice will keep water cold on a hot day.

12 Use two sections of the graph to explain what happens when you put a pot of cold water on a stove to boil. Specify which two sections you used.

13 Describe how you would separate a mixture of sugar, sand, and water.

14 The reaction of zinc metal with hydro-chloric acid produces zinc chloride and hydrogen gas. A student writes the following to represent the reaction.

$$Zn + HCl \rightarrow ZnCl_2 + H_2$$

Is the equation correct? Use conservation of mass to support your answer.

NEED EXTRA HELP?														
If You Missed Question...	1	2	3	4	5	6	7	8	9	10	11	12	13	14
Go to Lesson...	1	1	4	3	1	2	4	4	3	4	3	3	2	4

States of Matter

THE BIG IDEA What physical changes and energy changes occur as matter goes from one state to another?

Inquiry Liquid Glass?

When you look at this blob of molten glass, can you envision it as a beautiful vase? The solid glass was heated in a furnace until it formed a molten liquid. Air is blown through a pipe to make the glass hollow and give it form.

- Can you identify a solid, a liquid, and a gas in the photo?

- What physical changes and energy changes do you think occurred when the glass changed state?

Get Ready to Read

What do you think?

Before you read, decide if you agree or disagree with each of these statements. As you read this chapter, see if you change your mind about any of the statements.

1 Particles moving at the same speed make up all matter.

2 The particles in a solid do not move.

3 Particles of matter have both potential energy and kinetic energy.

4 When a solid melts, thermal energy is removed from the solid.

5 Changes in temperature and pressure affect gas behavior.

6 If the pressure on a gas increases, the volume of the gas also increases.

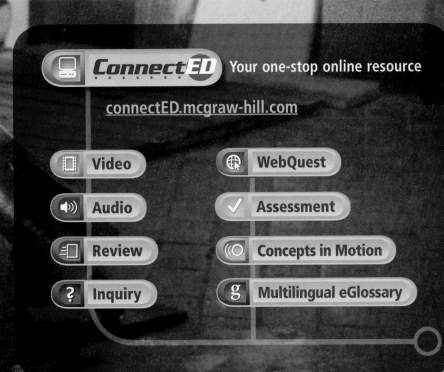

ConnectED Your one-stop online resource

connectED.mcgraw-hill.com

- Video
- Audio
- Review
- Inquiry
- WebQuest
- Assessment
- Concepts in Motion
- g Multilingual eGlossary

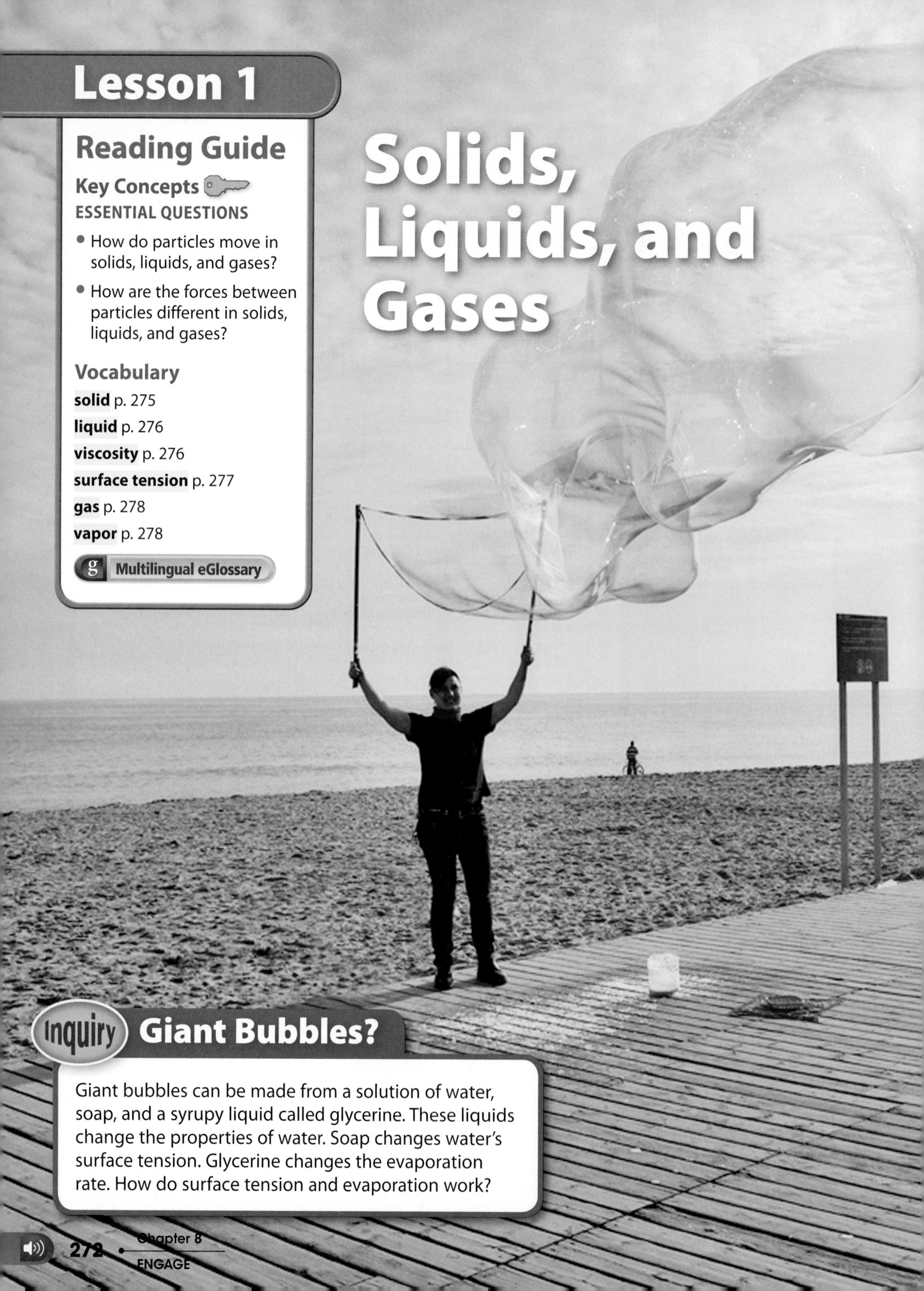

Reading Guide

Key Concepts 🔑
ESSENTIAL QUESTIONS

- How do particles move in solids, liquids, and gases?

- How are the forces between particles different in solids, liquids, and gases?

Vocabulary

solid p. 275

liquid p. 276

viscosity p. 276

surface tension p. 277

gas p. 278

vapor p. 278

g Multilingual eGlossary

Solids, Liquids, and Gases

Inquiry Giant Bubbles?

Giant bubbles can be made from a solution of water, soap, and a syrupy liquid called glycerine. These liquids change the properties of water. Soap changes water's surface tension. Glycerine changes the evaporation rate. How do surface tension and evaporation work?

How can you see particles in matter?

It's sometimes difficult to picture how tiny objects, such as the particles that make up matter, move. However, you can use other objects to model the movement of these particles.

1. Read and complete a lab safety form.

2. Place about 50 **copper pellets** into a **plastic petri dish.** Place the cover on the dish, and secure it with **tape.**

3. Hold the dish by the edges. Gently vibrate the dish from side to side no more than 1–2 mm. Observe the pellets. Record your observations in your Science Journal.

4. Repeat step 3, vibrating the dish less than 1 cm from side to side.

5. Repeat step 3, vibrating the dish 3–4 cm from side to side.

Think About This

1. If the pellets represent particles in matter, what do you think the shaking represents?

2. In which part of the experiment do you think the pellets were like a liquid? Explain.

3. 🔑 **Key Concept** If the pellets represent molecules of water, what do you think are the main differences among molecules of ice, water, and vapor?

Describing Matter

Take a closer look at the photo on the previous page. Do you see **matter?** The three most common forms, or states, of matter on Earth are solids, liquids, and gases. The giant bubble contains air, which is a mixture of gases. The ocean water and the soap mixture used to make the bubble are liquids. The sand, sign, and walkway are a few of the solids in the photo.

There is a fourth state of matter, plasma, that is not shown in this photo. Plasma is high-energy matter consisting of positively and negatively charged particles. Plasma is the most common state of matter in space. It also is in lightning flashes, fluorescent lighting, and stars, such as the Sun.

There are many ways to describe matter. You can describe the state, the color, the texture, and the odor of matter using your senses. You also can describe matter using measurements, such as mass, volume, and density. Mass is the amount of matter in an object. The units for mass are often grams (g) or kilograms (kg). Volume is the amount of space that a sample of matter occupies. The units for liquid volume are usually liters (L) or milliliters (mL). The units for solid volume are usually cubic centimeters (cm^3) or cubic meters (m^3). Density is the mass per unit volume of a substance. The units are usually g/cm^3 or g/mL. Density of a given substance remains constant, regardless of the size of the sample.

REVIEW VOCABULARY

matter
anything that takes up space and has mass

Particles in Motion

Have you ever wondered what makes something a solid, a liquid, or a gas? Two main factors that determine the state of matter are particle motion and particle forces.

Particles, such as atoms, ions, or molecules, moving in different ways make up all matter. The particles that make up some matter are close together and vibrate back and forth. In other types of matter, the particles are farther apart, move freely, and can spread out. Regardless of how close particles are to each other, they all move in random motion—movement in all directions and at different speeds. However, particles will move in straight lines until they collide with something. Collisions usually change the speed and direction of the particles' movements.

Forces Between Particles

Recall that atoms that make up matter contain positively charged protons and negatively charged electrons. There is a force of attractions between these oppositely charged particles, as shown in **Figure 1.**

You just read that the particles that make up matter move at all speeds and in all directions. If the motion of particles slows, the particles move closer together. This is because the attraction between them pulls them toward each other. Strong attractive forces hold particles close together. As the motion of particles increases, particles move farther apart. The attractive forces between particles get weaker. The spaces between them increase and the particles can slip past one another. As the motion of particles continues to increase, they move even farther apart. Eventually, the distance between particles is so great that there are little or no attractive forces between the particles. The particles move randomly and spread out. As you continue to read, you will learn how particle motion and particle forces determine whether matter is a solid, a liquid, or a gas.

FOLDABLES

Use a sheet of notebook paper to make a three-tab Foldable as shown. Record information about each state of matter under the tabs.

Solid

Liquid

Gas

Figure 1 The forces between particles of matter and the movement of particles determine the physical state of matter.

Concepts in Motion

Animation

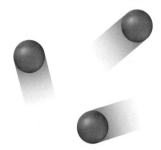

Particles move slowly and can only vibrate in place. Therefore, the attractive forces between particles are strong.

Particles move faster and slip past each other. The distance between particles increases. Therefore, the attractive forces between particles are weaker.

Particles move fast. The distance between the particles is great, and therefore, the attractive forces between particles are very weak.

Solids

If you had to describe a solid, what would you say? You might say, a **solid** *is matter that has a definite shape and a definite volume.* For example, if the skateboard in **Figure 2** moves from one location to another, the shape and volume of it do not change.

Particles in a Solid

Why doesn't a solid change its shape and volume? Notice in **Figure 2** how the particles in a solid are close together. The particles are very close to their neighboring particles. That's because the attractive forces between the particles are strong and hold them close together. The strong attractive forces and slow motion of the particles keep them tightly held in their positions. The particles simply vibrate back and forth in place. This arrangement gives solids a definite shape and volume.

 Key Concept Check Describe the movement of particles in a solid and the forces between them.

Types of Solids

All solids are not the same. For example, a diamond and a piece of charcoal don't look alike. However, they are both solids made of only carbon atoms. A diamond and a lump of charcoal both contain particles that strongly attract each other and vibrate in place. What makes them different is the arrangement of their particles. Notice in **Figure 3** that the arrangement of particles in a diamond is different from that in charcoal. A diamond is a crystalline solid. It has particles arranged in a specific, repeating order. Charcoal is an amorphous solid. It has particles arranged randomly. Different particle arrangements give these materials different properties. For example, a diamond is a hard material, and charcoal is a brittle material.

Reading Check What is the difference between crystalline and amorphous solids?

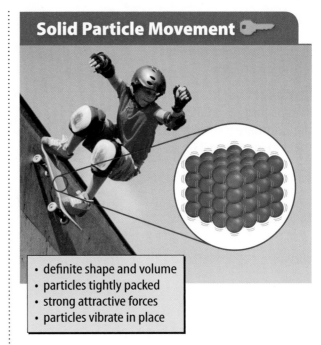

Solid Particle Movement 🔑

- definite shape and volume
- particles tightly packed
- strong attractive forces
- particles vibrate in place

▲ **Figure 2** The particles in a solid have strong attractive forces and vibrate in place.

Figure 3 Carbon is a solid that can have different particle arrangements. ▼

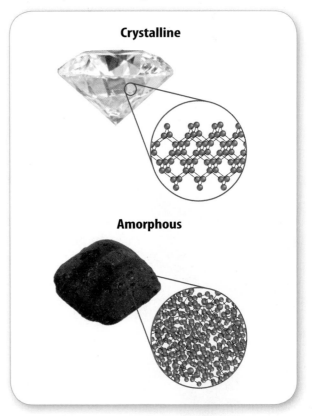

Crystalline

Amorphous

Figure 4 The motion of particles in a liquid causes the particles to move slightly farther apart. ▶

🔍 **Visual Check** How does the spacing among these particles compare to the particle spacing in **Figure 2?**

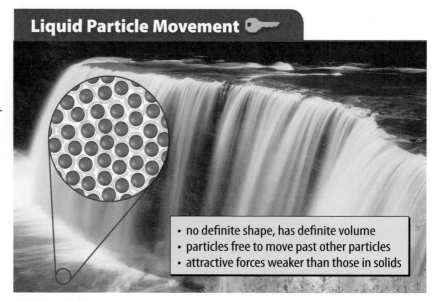

Liquid Particle Movement 🔑

- no definite shape, has definite volume
- particles free to move past other particles
- attractive forces weaker than those in solids

WORD ORIGIN ············

viscosity
from Latin *viscum*, means "sticky"
·················

Liquids

You have probably seen a waterfall, such as the one in **Figure 4.** Water is a liquid. *A* **liquid** *is matter with a definite volume but no definite shape.* Liquids flow and can take the shape of their containers. The container for this water is the riverbed.

Particles in a Liquid

How can liquids change their shape? The particle motion in the liquid state of a material is faster than the particle motion in the solid state. This increased particle motion causes the particles to move slightly farther apart. As the particles move farther apart, the attractive forces between the particles decrease. The weaker attractive forces allow particles to slip past one another. The weather forces also enable liquids to flow and take the shape of their containers.

Viscosity

If you have ever poured or dipped honey, as shown in **Figure 5,** you have experienced a liquid with a high viscosity. **Viscosity** (vihs KAW sih tee) *is a measurement of a liquid's resistance to flow.* Honey has high viscosity, while water has low viscosity. Viscosity is due to particle mass, particle shape, and the strength of the attraction between the particles of a liquid. In general, the stronger the attractive forces between particles, the higher the viscosity. For many liquids, viscosity decreases as the liquid becomes warmer. As a liquid becomes warmer, particles begin to move faster and the attractive forces between them get weaker. This allows particles to more easily slip past one another. The mass and shape of particles that make up a liquid also affect viscosity. Large particles or particles with complex shapes tend to move more slowly and have difficulty slipping past one another.

Figure 5 Honey has a high viscosity. ▼

Figure 6 The surface tension of water enables this spider to walk on the surface of a lake.

Surface Tension

How can the nursery web spider in **Figure 6** walk on water? Believe it or not, it is because of the interactions between molecules.

The blowout in **Figure 6** shows the attractive forces between water molecules. Water molecules below the surface are surrounded on all sides by other water molecules. Therefore, they have attractive forces, or pulls, in all directions. The attraction between similar molecules, such as water molecules, is called cohesion.

Water molecules at the surface of a liquid do not have liquid water molecules above them. As a result, they experience a greater downward pull, and the surface particles become tightly stretched like the head of a drum. Molecules at the surface of a liquid have **surface tension**, *the uneven forces acting on the particles on the surface of a liquid.* Surface tension allows a spider to walk on water. In general, the stronger the attractive forces between particles, the greater the surface tension of the liquid.

Recall the giant bubbles at the beginning of the chapter. The thin water-soap film surrounding the bubbles forms because of surface tension between the particles.

Key Concept Check Describe the movement of particles in a liquid and the forces between them.

Inquiry MiniLab

20 minutes

How can you make bubble films?

Have you ever observed surface tension? Which liquids have greater surface tension?

1. Read and complete a lab safety form.
2. Place about 100 mL of cool water in a **small bowl.** Lower a **wire bubble frame** into the bowl, and gently lift it. Use a **magnifying lens** to observe the edges of the frame. Write your observations in your Science Journal.
3. Add a full **dropper** of **liquid dishwashing soap** to the water. Stir with a **toothpick** until mixed. Lower the frame into the mixture and lift it out. Record your observations.
4. Use a toothpick to break the bubble film on one side of the thread. Observe.

Analyze and Conclude

1. **Recognize Cause and Effect** Explain what caused the thread to form an arc when half the bubble film broke.

2. **Key Concept** Explain why pure water doesn't form bubbles. What happens to the forces between water molecules when you add soap?

- no definite shape or volume
- particles far apart and move freely
- slight or weak attractive forces between particles

Figure 7 The particles in a gas are far apart, and there are little or no attractive forces between particles.

✓ Visual Check What are gas particles likely to hit as they move?

Gases

Look at the photograph in **Figure 7.** Where is the gas? *A* **gas** *is matter that has no definite volume and no definite shape.* It is not easy to identify the gas because you cannot see it. However, gas particles are inside and outside the inflatable balls. Air is a mixture of gases, including nitrogen, oxygen, argon, and carbon dioxide.

 Reading Check What is a gas, and what is another object that contains a gas?

Particles in a Gas

Why don't gases have definite volumes or definite shapes like solids and liquids? Compare the particles in **Figures 2, 4,** and **7.** Notice how the distance between particles differs. As the particles move faster, such as when matter goes from the solid state to the liquid state, the particles move farther apart. When the particles in matter move even faster, such as when matter goes from the liquid state to the gas state, the particles move even farther apart. When the distances between particles change, the attractive forces between the particles also change.

Forces Between Particles

As a type of matter goes from the solid state to the liquid state, the distance between the particles increases and the attractive forces between the particles decrease. When the same matter goes from the liquid state to the gas state, the particles are even farther apart and the attractive forces between the particles are weak or absent. As a result, the particles spread out to fill their container. Because gas particles lack attractive forces between particles, they have no definite shape or definite volume.

Vapor

Have you ever heard the term *vapor? The gas state of a substance that is normally a solid or a liquid at room temperature is called* **vapor.** For example, water is normally a liquid at room temperature. When it is in a gas state, such as in air, it is called water vapor. Other substances that can form a vapor are rubbing alcohol, iodine, mercury, and gasoline.

🔑 Key Concept Check How do particles move and interact in a gas?

Visual Summary

The particles that make up a solid can only vibrate in place. The particles are close together, and there are strong forces among them.

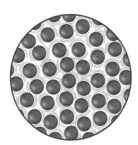

The particles that make up a liquid are far enough apart that particles can flow past other particles. The forces among these particles are weaker than those in a solid.

The particles that make up a gas are far apart. There is little or no attraction between the particles.

FOLDABLES

Use your lesson Foldable to review the lesson. Save your Foldable for the project at the end of the chapter.

What do you think NOW?

You first read the statements below at the beginning of the chapter.

1. Particles moving at the same speed make up all matter.

2. The particles in a solid do not move.

Did you change your mind about whether you agree or disagree with the statements? Rewrite any false statements to make them true.

Use Vocabulary

1 A measurement of how strongly particles attract one another at the surface of a liquid is _____.

2 **Define** *solid, liquid,* and *gas* in your own words.

3 A measurement of a liquid's resistance to flow is known as _____.

Understand Key Concepts

4 Which state of matter rarely is found on Earth?
- **A.** gas
- **B.** liquid
- **C.** plasma
- **D.** solid

5 **Compare** particle movement in solids, liquids, and gases.

6 **Compare** the forces between particles in a liquid and in a gas.

Interpret Graphics

7 **Explain** why the particles at the surface in the image below have surface tension while the particles below the surface do not.

8 **Summarize** Copy and fill in the graphic organizer to compare two types of solids.

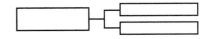

Critical Thinking

9 **Hypothesize** how you could change the viscosity of a cold liquid, and explain why your idea would work.

10 **Summarize** the relationship between the motion of particles and attractive forces between particles.

Freeze-Drying Foods

Have you noticed that the berries you find in some breakfast cereals are lightweight and dry—much different from the berries you get from the market or the garden?

Fresh fruit would spoil quickly if it were packaged in breakfast cereal, so fruits in cereals are often freeze-dried. When liquid is returned to the freeze-dried fruit, its physical properties more closely resemble fresh fruit. Freeze-drying, or lyophilization (lie ah fuh luh ZAY shun), is the process in which a solvent (usually water) is removed from a solid. During this process, a frozen solvent changes to a gas without going through the liquid state. Freeze-dried foods are lightweight and long-lasting. Astronauts have been using freeze-dried food during space travel since the 1960s.

How Freeze-Drying Works

❶ Machines called freeze-dryers are used to freeze-dry foods and other products. Fresh or cooked food is flash-frozen, changing moisture in the food to a solid.

❷ The frozen food is placed in a large vacuum chamber, where moisture is removed. Heat is applied to accelerate moisture removal. Condenser plates remove vaporized solvent from the chamber and convert the frozen food to a freeze-dried solid.

❸ Freeze-dried food is sealed in oxygen- and moisture-proof packages to ensure stability and freshness. When the food is rehydrated, it returns to its near-normal state of weight, color, and texture.

It's Your Turn

PREDICT/DISCOVER What kinds of products besides food are freeze-dried? Use library or internet resources to learn about other products that undergo the freeze-drying process. Discuss the benefits or drawbacks of freeze-drying.

Lesson 2

Reading Guide

Key Concepts
ESSENTIAL QUESTIONS

- How is temperature related to particle motion?
- How are temperature and thermal energy different?
- What happens to thermal energy when matter changes from one state to another?

Vocabulary

kinetic energy p. 282

temperature p. 282

thermal energy p. 283

vaporization p. 285

evaporation p. 286

condensation p. 286

sublimation p. 286

deposition p. 286

g **Multilingual eGlossary**

 Video

- **BrainPOP®**
- **What's Science Got to do With It?**

Changes in State

Inquiry **Spring Thaw?**

When you look at a snowman, you probably don't think about states of matter. However, water is one of the few substances that you frequently observe in three states of matter at Earth's temperatures. What energy changes are involved when matter changes state?

Do liquid particles move?

If you look at a glass of milk sitting on a table, it appears to have no motion. But appearances can be deceiving!

1. Read and complete a lab safety form.

2. Use a **dropper,** and place one drop of **2 percent milk** on a **glass slide.** Add a **cover slip.**

3. Place the slide on a **microscope** stage, and focus on low power. Focus on a single globule of fat in the milk. Observe the motion of the globule for several minutes. Record your observations in your Science Journal.

Think About This

1. Describe the motion of the fat globule.

2. What do you think caused the motion of the globule?

3. **Key Concept** What do you think would happen to the motion of the fat globule if you warmed the milk? Explain.

Kinetic and Potential Energy

When snow begins to melt after a snowstorm, all three states of water are present. The snow is a solid, the melted snow is a liquid, and the air above the snow and ice contains water vapor, a gas. What causes particles to change state?

Kinetic Energy

Recall that the particles that make up matter are in constant motion. These particles have **kinetic energy,** *the energy an object has due to its motion.* The faster particles move, the more kinetic energy they have. Within a given substance, such as water, particles in the solid state have the least amount of kinetic energy. This is because they only vibrate in place. Particles in the liquid state move faster than particles in the solid state. Therefore, they have more kinetic energy. Particles in the gaseous state move very quickly and have the most kinetic energy of particles of a given substance.

Temperature *is a measure of the average kinetic energy of all the particles in an object.* Within a given substance, a temperature increase means that the particles, on average, are moving at greater speeds, or have a greater average kinetic energy. For example, water molecules at 25°C are generally moving faster and have more kinetic energy than water molecules at 10°C.

Key Concept Check How is temperature related to particle motion?

Potential Energy

In addition to kinetic energy, particles have potential energy. Potential energy is stored energy due to the interactions between particles or objects. For example, when you pick up a ball and then let it go, the gravitational force between the ball and Earth causes the ball to fall toward Earth. Before you let the ball go, it has potential energy.

Potential energy typically increases when objects get farther apart and decreases when they get closer together. The basketball in the top part of **Figure 8** is farther off the ground than it is in the bottom part of the figure. The farther an object is from Earth's surface, the greater the gravitational potential energy. As the ball gets closer to the ground, the potential energy decreases.

You can think of the potential energy of particles in a similar way. The chemical potential energy is due to the position of the particles relative to other particles. The chemical potential energy of particles increases and decreases as the distances between particles increase or decrease. The particles in the top part of **Figure 8** are farther apart than the particles in the bottom part. The particles that are farther apart have greater chemical potential energy.

Thermal Energy

Thermal energy *is the total potential and kinetic energies of an object.* You can change an object's state of matter by adding or removing thermal energy. When you add thermal energy to an object, the particles either move faster (increased kinetic energy) or get farther apart (increased potential energy) or both. The opposite is true when you remove thermal energy from an object. If enough thermal energy is added or removed, a change of state can occur.

 Key Concept Check How do thermal energy and temperature differ?

Figure 8 The potential energy of the ball depends on the distance between the ball and Earth. The potential energy of particles in matter depends on the distances between the particles.

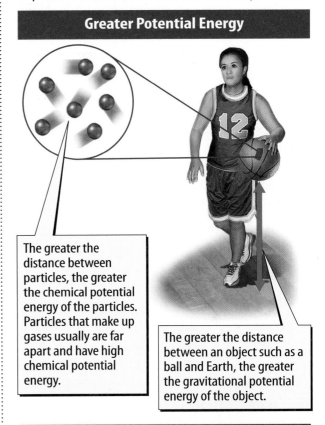

Greater Potential Energy

The greater the distance between particles, the greater the chemical potential energy of the particles. Particles that make up gases usually are far apart and have high chemical potential energy.

The greater the distance between an object such as a ball and Earth, the greater the gravitational potential energy of the object.

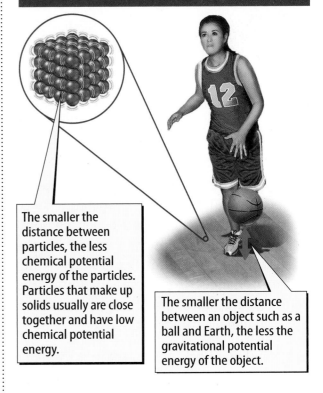

Less Potential Energy

The smaller the distance between particles, the less chemical potential energy of the particles. Particles that make up solids usually are close together and have low chemical potential energy.

The smaller the distance between an object such as a ball and Earth, the less the gravitational potential energy of the object.

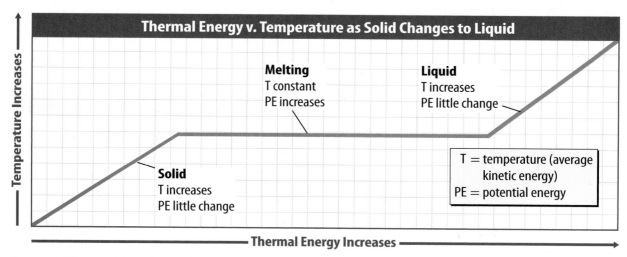

Thermal Energy v. Temperature as Solid Changes to Liquid

Temperature Increases →

Melting
T constant
PE increases

Liquid
T increases
PE little change

Solid
T increases
PE little change

T = temperature (average kinetic energy)
PE = potential energy

← Thermal Energy Increases →

Figure 9 Adding thermal energy to matter causes the particles that make up the matter to increase in kinetic energy, potential energy, or both.

Visual Check During melting, which factor remains constant?

Solid to Liquid or Liquid to Solid

When you drink a beverage from an aluminum can, do you recycle the can? Aluminum recycling is one example of a process that involves changing matter from one state to another by adding or removing thermal energy.

Melting

The first part of the recycling process involves melting aluminum cans. To change matter from a solid to a liquid, thermal energy must be added. The graph in **Figure 9** shows the relationship between increasing temperature and increasing thermal energy (potential energy + kinetic energy).

At first, both the thermal energy and the temperature increase. The temperature stops increasing when it reaches the melting point of the matter, the temperature at which the solid state changes to the liquid state. As aluminum changes from solid to liquid, the temperature does not change. However, energy changes still occur.

Reading Check What is added to matter to change it from a solid to a liquid?

Energy Changes

What happens when a solid reaches its melting point? Notice the line on the graph is horizontal. This means that the temperature, or average kinetic energy, stops increasing. However, the amount of thermal energy continues to increase. How is this possible?

Once a solid reaches its melting point, the average speed of particles does not change, but the distance between the particles does change. The particles move farther apart and potential energy increases. Once a solid completely melts, the addition of thermal energy will cause the kinetic energy of the particles to increase again, as shown by a temperature increase.

Freezing

After the aluminum melts, it is poured into molds to cool. As the aluminum cools, thermal energy leaves it. Freezing is a process that is the reverse of melting. The temperature at which matter changes from the liquid state to the solid state is its freezing point. To observe the temperature and thermal energy changes that occur to hot aluminum blocks, move from right to left on the graph in **Figure 9**.

During evaporation, a liquid vaporizes only at its surface.

During boiling, a liquid vaporizes at its surface and within the liquid.

Bubbles, or vaporized particles, rise to the top of the liquid and escape from the container.

Liquid to Gas or Gas to Liquid

When you heat water, do you ever notice how bubbles begin to form at the bottom and rise to the surface? The bubbles contain water vapor, a gas. *The change in state of a liquid into a gas is* **vaporization.** **Figure 10** shows two types of vaporization—evaporation and boiling.

Boiling

Vaporization that occurs within a liquid is called boiling. The temperature at which boiling occurs in a liquid is called its boiling point. In **Figure 11,** notice the energy changes that occur during this process. The kinetic energy of particles increases until the liquid reaches its boiling point.

At the boiling point, the potential energy of particles begins increasing. The particles move farther apart until the attractive forces no longer hold them together. At this point, the liquid changes to a gas. When boiling ends, if thermal energy continues to be added, the kinetic energy of the gas particles begins to increase again. Therefore, the temperature begins to increase again as shown on the graph.

▲ **Figure 10** 🔑 Boiling and evaporation are two kinds of vaporization.

✓ **Visual Check** Why doesn't the evaporation flask have bubbles below the surface?

🖳 **Review**

Personal Tutor

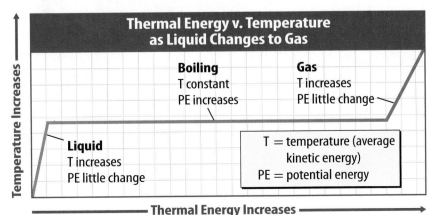

Thermal Energy v. Temperature as Liquid Changes to Gas

Temperature Increases →

Boiling
T constant
PE increases

Gas
T increases
PE little change

Liquid
T increases
PE little change

T = temperature (average kinetic energy)
PE = potential energy

— **Thermal Energy Increases** →

◀ **Figure 11** 🔑 When thermal energy is added to a liquid, kinetic energy and potential energy changes occur.

WORD ORIGIN

evaporation
from Latin *evaporare*, means
"disperse in steam or vapor"

Evaporation

Unlike boiling, **evaporation** *is vaporization that occurs only at the surface of a liquid.* Liquid in an open container will vaporize, or change to a gas, over time due to evaporation.

Condensation

Boiling and evaporation are processes that change a liquid to a gas. A reverse process also occurs. When a gas loses enough thermal energy, the gas changes to a liquid, or condenses. *The change of state from a gas to a liquid is called* **condensation.** Overnight, water vapor often condenses on blades of grass, forming dew.

Solid to Gas or Gas to Solid

Is it possible for a solid to become a gas without turning to a liquid first? Yes, in fact, dry ice does. Dry ice, as shown in **Figure 12,** is solid carbon dioxide. It turns immediately into a gas when thermal energy is added to it. The process is called sublimation. **Sublimation** *is the change of state from a solid to a gas without going through the liquid state.* As dry ice sublimes, it cools and condenses the water vapor in the surrounding air, creating a thick fog.

SCIENCE USE V. COMMON USE

deposition
Science Use the change of state of a gas to a solid without going through the liquid state

Common Use giving a legal testimony under oath

The opposite of sublimation is deposition. **Deposition** *is the change of state of a gas to a solid without going through the liquid state.* For deposition to occur, thermal energy has to be removed from the gas. You might see deposition in autumn when you wake up and there is frost on the grass. As water vapor loses thermal energy, it changes into a solid known as frost.

 Reading Check Why are sublimation and deposition unusual changes of state?

Figure 12 Dry ice sublimes—goes directly from the solid state to the gas state—when thermal energy is added. Frost is an example of the opposite process—deposition.

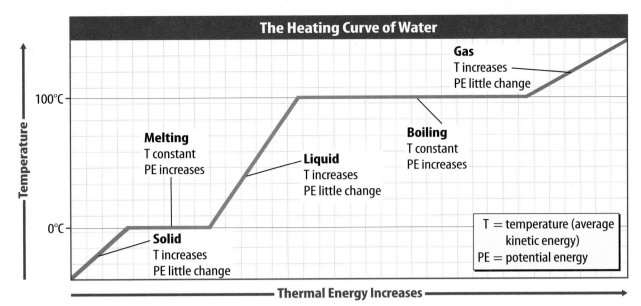

The Heating Curve of Water

Gas
T increases
PE little change

Melting
T constant
PE increases

Liquid
T increases
PE little change

Boiling
T constant
PE increases

100°C

Solid
T increases
PE little change

0°C

T = temperature (average kinetic energy)
PE = potential energy

Temperature

Thermal Energy Increases

States of Water

Water is the only substance that exists naturally as a solid, a liquid, and a gas within Earth's temperature range. To better understand the energy changes during a change in state, it is helpful to study the heating curve of water, as shown in **Figure 13.**

Adding Thermal Energy

Suppose you place a beaker of ice on a hot plate. The hot plate transfers thermal energy to the beaker and then to the ice. The temperature of the ice increases. Recall that this means the average kinetic energy of the water molecules increases.

At 0°C, the melting point of water, the water molecules vibrate so rapidly that they begin to move out of their places. At this point, added thermal energy only increases the distance between particles and decreases attractive forces—melting occurs. Once melting is complete, the average kinetic energy of the particles (temperature) begins to increase again as more thermal energy is added.

When water reaches 100°C, the boiling point, liquid water begins to change to water vapor. Again, kinetic energy is constant as vaporization occurs. When the change of state is complete, the kinetic energy of molecules increases once more, and so does the temperature.

 Key Concept Check Describe the changes in thermal energy as water goes from a solid to a liquid.

Removing Thermal Energy

The removal of thermal energy is the reverse of the process shown in **Figure 13.** Cooling water vapor changes the gas to a liquid. Cooling the water further changes it to ice.

Figure 13 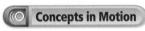 Water undergoes energy changes and state changes as thermal energy is added and removed.

Concepts in Motion

Animation

Fold a sheet of notebook paper to make a four-tab Foldable as shown. Label the tabs, define the terms, and record what you learn about each term under the tabs.

Vaporization
Boiling Evaporation

Condensation

Sublimation

Deposition

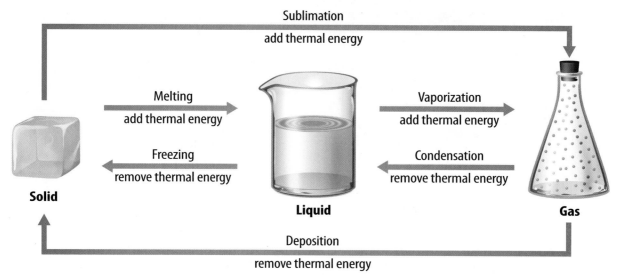

Sublimation
add thermal energy

Melting
add thermal energy

Vaporization
add thermal energy

Freezing
remove thermal energy

Condensation
remove thermal energy

Solid

Liquid

Gas

Deposition
remove thermal energy

Figure 14 🔑 For a change of state to occur, thermal energy must move into or out of matter.

Conservation of Mass and Energy

The diagram in **Figure 14** illustrates the energy changes that occur as thermal energy is added or removed from matter. Notice that opposite processes, melting and freezing and vaporization and condensation, are shown. When matter changes state, matter and energy are always conserved.

When water vaporizes, it appears to disappear. If the invisible gas is captured and its mass added to the remaining mass of the liquid, you would see that matter is conserved. This is also true for energy. Surrounding matter, such as air, often absorbs thermal energy. If you measured all the thermal energy, you would find that energy is conserved.

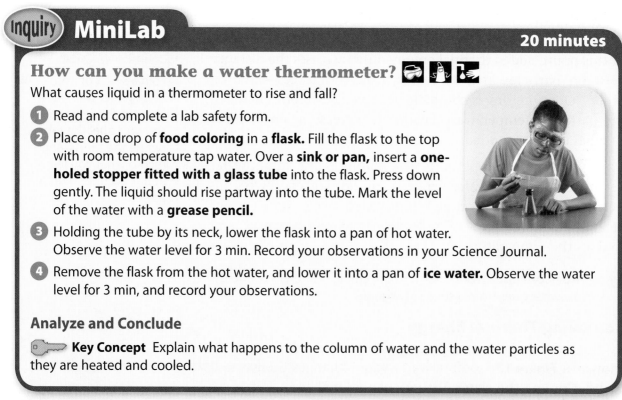

Inquiry MiniLab

20 minutes

How can you make a water thermometer? 🥽 🦺 🧤

What causes liquid in a thermometer to rise and fall?

1. Read and complete a lab safety form.

2. Place one drop of **food coloring** in a **flask.** Fill the flask to the top with room temperature tap water. Over a **sink or pan,** insert a **one-holed stopper fitted with a glass tube** into the flask. Press down gently. The liquid should rise partway into the tube. Mark the level of the water with a **grease pencil.**

3. Holding the tube by its neck, lower the flask into a pan of hot water. Observe the water level for 3 min. Record your observations in your Science Journal.

4. Remove the flask from the hot water, and lower it into a pan of **ice water.** Observe the water level for 3 min, and record your observations.

Analyze and Conclude

🔑 **Key Concept** Explain what happens to the column of water and the water particles as they are heated and cooled.

Lesson 2 Review

Visual Summary

All matter has thermal energy. Thermal energy is the sum of potential and kinetic energy.

When thermal energy is added to a liquid, vaporization can occur.

When enough thermal energy is removed from matter, a change of state can occur.

FOLDABLES

Use your lesson Foldable to review the lesson. Save your Foldable for the project at the end of the chapter.

What do you think NOW?

You first read the statements below at the beginning of the chapter.

3. Particles of matter have both potential energy and kinetic energy.

4. When a solid melts, thermal energy is removed from the solid.

Did you change your mind about whether you agree or disagree with the statements? Rewrite any false statements to make them true.

Use Vocabulary

1. The measure of average kinetic energy of the particles in a material is _____.

2. **Define** *kinetic energy* and *thermal energy* in your own words.

3. The change of a liquid into a gas is known as _____.

Understand Key Concepts

4. The process that is opposite of condensation is known as
   A. deposition.   C. melting.
   B. freezing.   D. vaporization.

5. **Explain** how temperature and particle motion are related.

6. **Describe** the relationship between temperature and thermal energy.

7. **Generalize** the changes in thermal energy when matter increases in temperature and then changes state.

Interpret Graphics

8. **Describe** what is occurring below.

9. **Summarize** Copy and fill in the graphic organizer below to identify the two types of vaporization that can occur in matter.

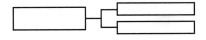

Critical Thinking

10. **Summarize** the energy and state changes that occur when freezing rain falls and solidifies on a wire fence.

11. **Compare** the amount of thermal energy needed to melt a solid and the amount of thermal energy needed to freeze the same liquid.

How does dissolving substances in water change its freezing point?

Materials

triple-beam balance

beaker

foam cup

50-mL graduated cylinder

distilled water

Also needed:
ice-salt slush, test tubes, thermometers

Safety

You know that when thermal energy is removed from a liquid, the particles move more slowly. At the freezing point, the particles move so slowly that the attractive forces pull them together to form a solid. What happens if the water contains particles of another substance, such as salt? You will form a hypothesis and test the hypothesis to find out.

Learn It

To **form a hypothesis** is to propose a possible explanation for an observation that is testable by a scientific investigation. You **test the hypothesis** by conducting a scientific investigation to see whether the hypothesis is supported.

Try It

1. Read and complete a lab safety form.

2. Form a hypothesis that answers the question in the title of the lab. Write your hypothesis in your Science Journal.

3. Copy the data table in your Science Journal.

4. Use a triple-beam balance to measure 5 g of table salt (NaCl). Dissolve the 5 g of table salt in 50 mL of distilled water.

5. Place 40 mL of distilled water in one large test tube. Place 40 mL of the salt-water mixture in a second large test tube.

6. Measure and record the temperature of the liquids in each test tube.

7. Place both test tubes into a large foam cup filled with crushed ice-salt slush. Gently rotate the thermometers in the test tubes. Record the temperature in each test tube every minute until the temperature remains the same for several minutes.

Apply It

8. How does the data tell you when the freezing point of the liquid has been reached?

9. Was your hypothesis supported? Why or why not?

10. 🔑 **Key Concept** Explain your observations in terms of how temperature affects particle motion and how a liquid changes to a solid.

Water	Time (min)	0	1	2	3	4	5	6	7	8
	Temperature (°C)									
Salt water	Time (min)	0	1	2	3	4	5	6	7	8
	Temperature (°C)									

Lesson 3

The Behavior of Gases

Reading Guide

Key Concepts 🔑
ESSENTIAL QUESTIONS

- How does the kinetic molecular theory describe the behavior of a gas?
- How are temperature, pressure, and volume related in Boyle's law?
- How is Boyle's law different from Charles's law?

Vocabulary

kinetic molecular theory p. 292

pressure p. 293

Boyle's law p. 294

Charles's law p. 295

 Multilingual eGlossary

 Video

What's Science Got to do With It?

(inquiry) Survival Gear?

Why do some pilots wear oxygen masks? Planes fly at high altitudes where the atmosphere has a lower pressure and gas molecules are less concentrated. If the pressure is not adjusted inside the airplane, a pilot must wear an oxygen mask to inhale enough oxygen to keep the body functioning.

Inquiry Launch Lab

15 minutes

Are volume and pressure of a gas related?

Pressure affects gases differently than it affects solids and liquids. How do pressure changes affect the volume of a gas?

1. Read and complete a lab safety form.
2. Stretch and blow up a **small balloon** several times.
3. Finally, blow up the balloon to a diameter of about 5 cm. Twist the neck, and stretch the mouth of the balloon over the opening of a **plastic bottle. Tape** the neck of the balloon to the bottle.
4. Squeeze and release the bottle several times while observing the balloon. Record your observations in your Science Journal.

Think About This

1. Why doesn't the balloon deflate when you attach it to the bottle?

2. What caused the balloon to inflate when you squeezed the bottle?

3. 🔑 **Key Concept** Using this lab as a reference, do you think pressure and volume of a gas are related? Explain.

Understanding Gas Behavior

Pilots do not worry as much about solids and liquids at high altitudes as they do gases. That is because gases behave differently than solids and liquids. Changes in temperature, pressure, and volume affect the behavior of gases more than they affect solids and liquids.

The explanation of particle behavior in solids, liquids, and gases is based on the kinetic molecular theory. The **kinetic molecular theory** *is an explanation of how particles in matter behave.* Some basic ideas in this theory are

- small particles make up all matter;

- these particles are in constant, random motion;

- the particles collide with other particles, other objects, and the walls of their container;

- when particles collide, no energy is lost.

You have read about most of these, but the last two statements are very important in explaining how gases behave.

🔑 **Key Concept Check** How does the kinetic molecular theory describe the behavior of a gas?

ACADEMIC VOCABULARY

theory
(noun) an explanation of things or events that is based on knowledge gained from many observations and investigations

Greatest volume, least pressure	Less volume, more pressure	Least volume, most pressure

Figure 15 🔑 As pressure increases, the volume of the gas decreases.

What is pressure?

Particles in gases move constantly. As a result of this movement, gas particles constantly collide with other particles and their container. When particles collide with their container, pressure results. **Pressure** *is the amount of force applied per unit of area.* For example, gas in a cylinder, as shown in **Figure 15,** might contain trillions of gas particles. These particles exert forces on the cylinder each time they strike it. When a weight is added to the plunger, the plunger moves down, compressing the gas in the cylinder. With less space to move around, the particles that make up the gas collide with each other more frequently, causing an increase in pressure. The more the particles are compressed, the more often they collide, increasing the pressure.

Pressure and Volume

Figure 15 also shows the relationship between pressure and volume of gas at a constant temperature. What happens to pressure if the volume of a container changes? Notice that when the volume is greater, the particles have more room to move. This additional space results in fewer collisions within the cylinder, and pressure is less. The gas particles in the middle cylinder have even less volume and more pressure. In the cylinder on the right, the pressure is greater because the volume is less. The particles collide with the container more frequently. Because of the greater number of collisions within the container, pressure is greater.

WORD ORIGIN

pressure
from Latin *pressura*, means "to press"

FOLDABLES®

Fold a sheet of notebook paper to make a three-tab Foldable and label as shown. Use your Foldable to compare two important gas laws.

Solve Equations

Boyle's law can be stated by the equation

$$V_2 = \frac{P_1 V_1}{P_2}$$

P_1 and V_1 represent the pressure and volume before a change. P_2 and V_2 are the pressure and volume after a change. Pressure is often measured in kilopascals (kPa). For example, what is the final volume of a gas with an initial volume of 50.0 mL if the pressure increases from 600.0 kPa to 900.0 kPa?

1. Replace the terms in the equation with the actual values.

$$V_2 = \frac{(600.0 \text{ kPa})(50.0 \text{ mL})}{(900.0 \text{ kPa})}$$

2. Cancel units, multiply, and then divide.

$$V_2 = \frac{(600.0 \text{ kPa})(50.0 \text{ mL})}{(900.0 \text{ kPa})}$$

$$V_2 = 33.3 \text{ mL}$$

Practice

What is the final volume of a gas with an initial volume of 100.0 mL if the pressure decreases from 500.0 kPa to 250.0 kPa?

 Review

- Math Practice
- Personal Tutor

Boyle's Law

You read that the pressure and volume of a gas are related. Robert Boyle (1627–1691), a British scientist, was the first to describe this property of gases. **Boyle's law** *states that pressure of a gas increases if the volume decreases and pressure of a gas decreases if the volume increases, when temperature is constant.* This law can be expressed mathematically as shown to the left.

 Key Concept Check What is the relationship between pressure and volume of a gas if temperature is constant?

Boyle's Law in Action

You have probably felt Boyle's law in action if you have ever traveled in an airplane. While on the ground, the air pressure inside your middle ear and the pressure of the air surrounding you are equal. As the airplane takes off and begins to increase in altitude, the air pressure of the surrounding air decreases. However, the air pressure inside your middle ear does not decrease. The trapped air in your middle ear increases in volume, which can cause pain. These pressure changes also occur when the plane is landing. You can equalize this pressure difference by yawning or chewing gum.

Graphing Boyle's Law

This relationship is shown in the graph in **Figure 16.** Pressure is on the x-axis, and volume is on the y-axis. Notice that the line decreases in value from left to right. This shows that as the pressure of a gas increases, the volume of the gas decreases.

Figure 16 The graph shows that as pressure increases, volume decreases. This is true only if the temperature of the gas is constant.

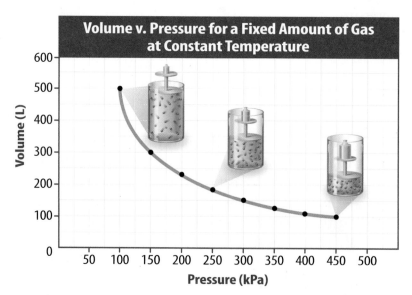

Concepts in Motion

Animation

Higher temperature, greater volume

Lower temperature, less volume

Figure 17 As the temperature of a gas increases, the kinetic energy of the particles increases. The particles move farther apart, and volume increases.

Temperature and Volume

Pressure and volume changes are not the only factors that affect gas behavior. Changing the temperature of a gas also affects its behavior, as shown in **Figure 17.** The gas in the cylinder on the left has a low temperature. The average kinetic energy of the particles is low, and they move closer together. The volume of the gas is less. When thermal energy is added to the cylinder, the gas particles move faster and spread farther apart. This increases the pressure from gas particles, which push up the plunger. This increases the volume of the container.

Charles's Law

Jacque Charles (1746–1823) was a French scientist who described the relationship between temperature and volume of a gas. **Charles's law** states that the volume of a gas increases with increasing temperature, if the pressure is constant. Charles's practical experience with gases was most likely the result of his interest in balloons. Charles and his colleague were the first to pilot and fly a hydrogen-filled balloon in 1783.

Key Concept Check How is Boyle's law different from Charles's law?

Inquiry MiniLab **20 minutes**

How does temperature affect the volume?

You can observe Charles's law in action using a few lab supplies.

1. Read and complete a lab safety form.

2. Stretch and blow up a **small balloon** several times.

3. Finally, blow up the balloon to a diameter of about 5 cm. Twist the neck and stretch the mouth of the balloon over the opening of an **ovenproof flask.**

4. Place the flask on a cold **hot plate.** Turn on the hot plate to low, and gradually heat the flask. Record your observations in your Science Journal.

5. ⚠ Use **tongs** to remove the flask from the hot plate. Allow the flask to cool for 5 min. Record your observations.

6. Place the flask in a **bowl of ice water.** Record your observations.

Analyze and Conclude

Key Concept What is the effect of temperature changes on the volume of a gas?

Charles's Law in Action

You have probably seen Charles's law in action if you have ever taken a balloon outside on a cold winter day. Why does a balloon appear slightly deflated when you take it from a warm place to a cold place? When the balloon is in cold air, the temperature of the gas inside the balloon decreases. Recall that a decrease in temperature is a decrease in the average kinetic energy of particles. As a result, the gas particles slow down and begin to get closer together. Fewer particles hit the inside of the balloon. The balloon appears partially deflated. If the balloon is returned to a warm place, the kinetic energy of the particles increases. More particles hit the inside of the balloon and push it out. The volume increases.

Reading Check What happens when you warm a balloon?

Graphing Charles's Law

The relationship described in Charles's law is shown in the graph of several gases in **Figure 18.** Temperature is on the *x*-axis and volume is on the *y*-axis. Notice that the lines are straight and represent increasing values. Each line in the graph is extrapolated to −273°C. *Extrapolated* means the graph is extended beyond the observed data points. This temperature also is referred to as 0 K (kelvin), or absolute zero. This temperature is theoretically the lowest possible temperature of matter. At absolute zero, all particles are at the lowest possible energy state and do not move. The particles contain a minimal amount of thermal energy (potential energy + kinetic energy).

Key Concept Check Which factors must be constant in Boyle's law and in Charles's law?

Figure 18 The volume of a gas increases when the temperature increases at constant pressure.

Visual Check What do the dashed lines mean?

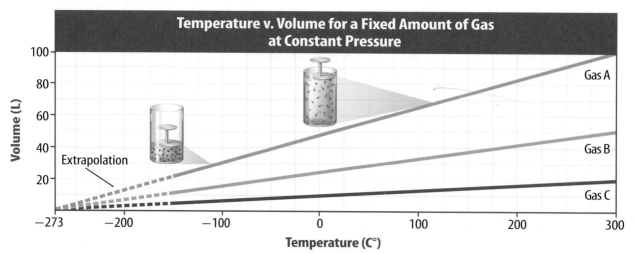

Temperature v. Volume for a Fixed Amount of Gas at Constant Pressure

Lesson 3 Review

Visual Summary

The explanation of particle behavior in solids, liquids, and gases is based on the kinetic molecular theory.

As volume of a gas decreases, the pressure increases when at constant temperature.

At constant pressure, as the temperature of a gas increases, the volume also increases.

FOLDABLES®

Use your lesson Foldable to review the lesson. Save your Foldable for the project at the end of the chapter.

What do you think NOW?

You first read the statements below at the beginning of the chapter.

5. Changes in temperature and pressure affect gas behavior.

6. If the pressure on a gas increases, the volume of the gas also increases.

Did you change your mind about whether you agree or disagree with the statements? Rewrite any false statements to make them true.

Use Vocabulary

1 **List** the basic ideas of the kinetic molecular theory.

2 _____ is force applied per unit area.

Understand Key Concepts

3 Which is held constant when a gas obeys Boyle's law?
   A. motion
   B. pressure
   C. temperature
   D. volume

4 **Describe** how the kinetic molecular theory explains the behavior of a gas.

5 **Contrast** Charles's law with Boyle's law.

6 **Explain** how temperature, pressure, and volume are related in Boyle's law.

Interpret Graphics

7 **Explain** what happens to the particles to the right when more weights are added.

8 **Identify** Copy and fill in the graphic organizer below to list three factors that affect gas behavior.

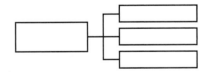

Critical Thinking

9 **Describe** what would happen to the pressure of a gas if the volume of the gas doubles while at a constant temperature.

Math Skills ×÷+   Review
——— Math Practice ———

10 **Calculate** The pressure on 400 mL of a gas is raised from 20.5 kPa to 80.5 kPa. What is the final volume of the gas?

Materials

triple-beam balance

50-mL graduated cylinders

beakers

test tubes

thermometers

distilled water

Also needed:
ice, salt

Safety

Design an Experiment to Collect Data

In this chapter, you have learned about the relationship between the motion of particles in matter and change of state. How might you use your knowledge of particles in real life? Suppose that you work for a state highway department in a cold climate. Your job is to test three products. You must determine which is the most effective in melting existing ice, the best at keeping melted ice from refreezing, and the best product to buy.

Question

How can you compare the products? What might make one product better than another? Consider how you can describe and compare the effect of each product on both existing ice and the freezing point of water. Think about controls, variables, and the equipment you have available.

Procedure

1. Read and complete a lab safety form.

2. In your Science Journal, write a set of procedures you will use to answer your questions. Include the materials and steps you will use to test the effect of each product on existing ice and on the freezing point of water. How will you record your data? Draw any data tables, such as the example below, that you might need. Have your teacher approve your procedures.

Distilled Water	Time (min)	0	1	2	3	4	5	6	7	8
	Temperature (°C)									
Product A	Time (min)	0	1	2	3	4	5	6	7	8
	Temperature (°C)									
Product B	Time (min)	0	1	2	3	4	5	6	7	8
	Temperature (°C)									
Product C	Time (min)	0	1	2	3	4	5	6	7	8
	Temperature (°C)									

3. Begin by observing and recording your observations on how each product affects ice. Does it make ice melt or melt faster?

4. Test the effect of each product on the freezing point of water. Think about how you will ensure that each product is tested in the same way.

5. Add any additional tests you think you might need to make your recommendation.

3

Analyze and Conclude

6 **Analyze the data** you have collected. Which product was most effective in melting existing ice? How do you know?

7 **Determine** which product was most effective in lowering the freezing point of water.

8 **Draw or make a model** to show the effect of dissolved solids on water molecules.

9 **Recognize Cause and Effect** In terms of particles, what causes dissolved solids to lower the freezing point of water?

10 **Draw Conclusions** In terms of particles, why are some substances more effective than others in lowering the freezing point of water?

11 🔵 **The Big Idea** Why is the kinetic molecular theory important in understanding how and why matter changes state?

Communicate Your Results

You are to present your recommendations to the road commissioners. Create a graphic presentation that clearly displays your results and justifies your recommendations about which product to buy.

 **Extension**

In some states, road crews spray liquid deicer on the roads. If your teacher approves, you may enjoy testing liquids, such as alcohol, corn syrup, or salad oil.

Lab Tips

☑ To ensure fair testing, add the same mass of each product to the ice cubes at the same time.

☑ Be sure to add the same mass of each solid to the same volume of water. About 1 g of solid in 10 mL of water is a good ratio.

☑ Keep adding crushed ice/salt slush to the cup so that the liquid in the test tubes remains below the surface.

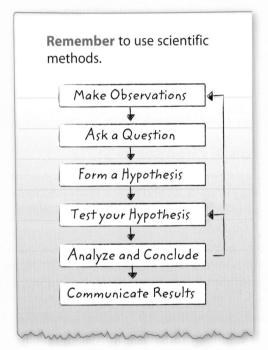

Remember to use scientific methods.

Make Observations
↓
Ask a Question
↓
Form a Hypothesis
↓
Test your Hypothesis
↓
Analyze and Conclude
↓
Communicate Results

Chapter 8 Study Guide

THE BIG IDEA As matter changes from one state to another, the distances and the forces between the particles change, and the amount of thermal energy in the matter changes.

Key Concepts Summary 🔑

Lesson 1: Solids, Liquids, and Gases

- Particles vibrate in **solids.** They move faster in **liquids** and even faster in **gases.**
- The force of attraction among particles decreases as matter goes from a solid, to a liquid, and finally to a gas.

Solid   Liquid   Gas

Lesson 2: Changes in State

- Because **temperature** is defined as the average **kinetic energy** of particles and kinetic energy depends on particle motion, temperature is directly related to particle motion.
- **Thermal energy** includes both the kinetic energy and the potential energy of particles in matter. However, temperature is only the average kinetic energy of particles in matter.
- Thermal energy must be added or removed from matter for a change of state to occur.

Lesson 3: The Behavior of Gases

- The **kinetic molecular theory** states basic assumptions that are used to describe particles and their interactions in gases and other states of matter.
- **Pressure** of a gas increases if the volume decreases, and pressure of a gas decreases if the volume increases, when temperature is constant.
- **Boyle's law** describes the behavior of a gas when pressure and volume change at constant temperature. **Charles's law** describes the behavior of a gas when temperature and volume change, and pressure is constant.

Vocabulary

solid p. 275
liquid p. 276
viscosity p. 276
surface tension p. 277
gas p. 278
vapor p. 278

kinetic energy p. 282
temperature p. 282
thermal energy p. 283
vaporization p. 285
evaporation p. 286
condensation p. 286
sublimation p. 286
deposition p. 286

kinetic molecular theory p. 292
pressure p. 293
Boyle's law p. 294
Charles's law p. 295

FOLDABLES® Chapter Project

Assemble your lesson Foldables as shown to make a Chapter Project. Use the project to review what you have learned in this chapter.

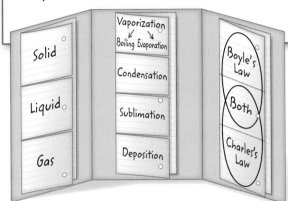

Use Vocabulary

Replace the underlined word with the correct term.

1 Matter with a definite shape and a definite volume is known as a <u>gas</u>.

2 <u>Surface tension</u> is a measure of a liquid's resistance to flow.

3 The gas state of a substance that is normally a solid or a liquid at room temperature is a <u>pressure</u>.

4 <u>Boiling</u> is vaporization that occurs at the surface of a liquid.

5 <u>Boyle's law</u> is an explanation of how particles in matter behave.

6 When graphing a gas obeying <u>Boyle's law</u>, the line will be a straight line with a positive slope.

Link Vocabulary and Key Concepts

◖◖ **Concepts in Motion**   Interactive Concept Map

Copy this concept map, and then use vocabulary terms from the previous page to complete the concept map.

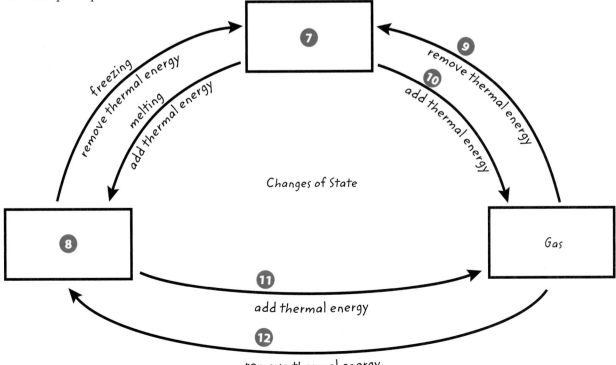

Understand Key Concepts

1. What would happen if you tried to squeeze a gas into a smaller container?
   A. The attractive forces between the particles would increase.
   B. The force of the particles would prevent you from doing it.
   C. The particles would have fewer collisions with the container.
   D. The repulsive forces of the particles would pull on the container.

2. Which type of motion in the figure below best represents the movement of gas particles?

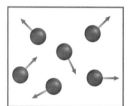 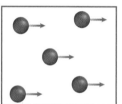

Motion 1   Motion 2

Motion 3   Motion 4

   A. motion 1
   B. motion 2
   C. motion 3
   D. motion 4

3. A pile of snow slowly disappears into the air, even though the temperature remains below freezing. Which process explains this?
   A. condensation
   B. deposition
   C. evaporation
   D. sublimation

4. Which unit is a density unit?
   A. cm^3
   B. cm^3/g
   C. g
   D. g/cm^3

5. Which is a form of vaporization?
   A. condensation
   B. evaporation
   C. freezing
   D. melting

6. When a needle is placed on the surface of water, it floats. Which idea best explains why this happens?
   A. Boyle's law
   B. molecular theory
   C. surface tension
   D. viscosity theory

7. In which material would the particles be most closely spaced?
   A. air
   B. brick
   C. syrup
   D. water

Use the graph below to answer questions 8 and 9.

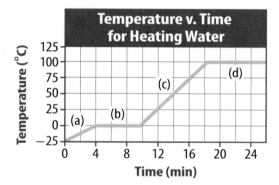

8. Which area of the graph above shows melting of a solid?
   A. a
   B. b
   C. c
   D. d

9. Which area or areas of the graph above shows a change in the potential energy of the particles?
   A. a
   B. a and c
   C. b and d
   D. c

Chapter Review

✓ **Assessment**

Online Test Practice

Critical Thinking

10 **Explain** how the distances between particles in a solid, a liquid, and a gas help determine the densities of each.

11 **Describe** what would happen to the volume of a balloon if it were submerged in hot water.

12 **Assess** The particles of an unknown liquid have very weak attractions for other particles in the liquid. Would you expect the liquid to have a high or low viscosity? Explain your answer.

13 **Rank** these liquids from highest to lowest viscosity: honey, rubbing alcohol, and ketchup.

14 **Evaluate** Each beaker below contains the same amount of water. The thermometers show the temperature in each beaker. Explain the kinetic energy differences in each beaker.

15 **Summarize** A glass with a few milliliters of water is placed on a counter. No one touches the glass. Explain what happens to the water after a few days.

Writing in Science

16 **Write** a paragraph that describes how you could determine the melting point of a substance from its heating or cooling curve.

REVIEW THE BIG IDEA

17 During springtime in Alaska, frozen rivers thaw and boats can navigate the rivers again. What physical changes and energy changes occur to the ice molecules when ice changes to water? Explain the process in which water in the river changes to water vapor.

18 In the photo below, explain how the average kinetic energy of the particles changes as the molten glass cools. What instrument could you use to verify the change in the average kinetic energy of the particles?

Math Skills

Review

Math Practice

Solve Equations

19 The pressure on 1 L of a gas at a pressure of 600 kPa is lowered to 200 kPa. What is the final volume of the gas?

20 A gas has a volume of 30 mL at a pressure of 5000 kPa. What is the volume of the gas if the pressure is lowered to 1,250 kPa?

25°C   75°C

Standardized Test Practice

Record your answers on the answer sheet provided by your teacher or on a separate sheet of paper.

Multiple Choice

1 Which property applies to matter that consists of particles vibrating in place?

  A has a definite shape

  B takes the shape of the container

  C flows easily at room temperature

  D particles far apart

Use the figure below to answer questions 2 and 3.

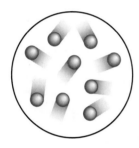

2 Which state of matter is represented above?

  A amorphous solid

  B crystalline solid

  C gas

  D liquid

3 Which best describes the attractive forces between particles shown in the figure?

  A The attractive forces keep the particles vibrating in place.

  B The particles hardly are affected by the attractive forces.

  C The attractive forces keep the particles close together but still allow movement.

  D The particles are locked in their positions because of the attractive forces between them.

4 What happens to matter as its temperature increases?

  A The average kinetic energy of its particles decreases.

  B The average thermal energy of its particles decreases.

  C The particles gain kinetic energy.

  D The particles lose potential energy.

Use the figure to answer question 5.

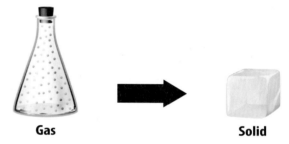

Gas   Solid

5 Which process is represented in the figure?

  A deposition

  B freezing

  C sublimation

  D vaporization

6 Which is a fundamental assumption of the kinetic molecular theory?

  A All atoms are composed of subatomic particles.

  B The particles of matter move in predictable paths.

  C No energy is lost when particles collide with one another.

  D Particles of matter never come into contact with one another.

7 Which is true of the thermal energy of particles?

  A Thermal energy includes the potential and the kinetic energy of the particles.

  B Thermal energy is the same as the average kinetic energy of the particles.

  C Thermal energy is the same as the potential energy of particles.

  D Thermal energy is the same as the temperature of the particles.

Use the graph below to answer question 8.

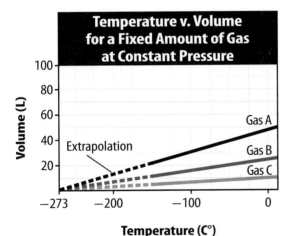

8 Which relationship is shown in the graph?

  A Boyle's law

  B Charles's law

  C kinetic molecular theory

  D definition of thermal energy

Constructed Response

9 Some people say that something that does not move very quickly is "as slow as molasses in winter." What property of molasses is described by the saying? Based on the saying, how do you think this property changes with temperature?

Use the graph to answer questions 10 and 11.

A scientist measured the temperature of a sample of frozen mercury as thermal energy is added to the sample. The graph below shows the results.

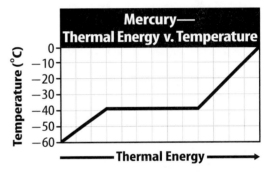

10 At what temperature does mercury melt? How do you know?

11 Describe the motion and arrangement of mercury atoms while the temperature is constant.

12 Atmospheric pressure is greater at the base of a mountain than at its peak. A hiker drinks from a water bottle at the top of a mountain. The bottle is capped tightly. At the base of the mountain, the water bottle has collapsed slightly. What happened to the gas inside the bottle? Assume constant temperature. Explain.

NEED EXTRA HELP?												
If You Missed Question...	1	2	3	4	5	6	7	8	9	10	11	12
Go to Lesson...	1	1	1	2	2	3	2	3	1	1	2	3

Student Resources

These resources are designed to help you achieve success in science. You will find useful information on laboratory safety, math skills, and science skills. In addition, science reference materials are found in the Reference Handbook. You'll find the information you need to learn and sharpen your skills in these resources.

Table of Contents

Science Skill Handbook SR-2

Scientific Methods **SR-2**
Identify a Question.. SR-2
Gather and Organize Information............................... SR-2
Form a Hypothesis SR-5
Test the Hypothesis SR-6
Collect Data SR-6
Analyze the Data.. SR-9
Draw Conclustions SR-10
Communicate.. SR-10
Safety Symbols **SR-11**
Safety in the Science Laboratory......................... **SR-12**
General Safety Rules SR-12
Prevent Accidents.. SR-12
Laboratory Work SR-13
Emergencies.. SR-13

Math Skill Handbook SR-14

Math Review.. **SR-14**
Use Fractions SR-14
Use Ratios SR-17
Use Decimals SR-17
Use Proportions.. SR-18
Use Percentages SR-19
Solve One-Step Equations..................................... SR-19
Use Statistics... SR-20
Use Geometry SR-21
Science Application **SR-24**
Measure in SI SR-24
Dimensional Analysis... SR-24
Precision and Significant Digits............................. SR-26
Scientific Notation SR-26
Make and Use Graphs SR-27

Foldables Handbook SR-29

Reference Handbook SR-40
Periodic Table of the Elements............................... SR-40

Glossary G-2

Index I-2

Credits C-2

Scientific Methods

Scientists use an orderly approach called the scientific method to solve problems. This includes organizing and recording data so others can understand them. Scientists use many variations in this method when they solve problems.

Identify a Question

The first step in a scientific investigation or experiment is to identify a question to be answered or a problem to be solved. For example, you might ask which gasoline is the most efficient.

Gather and Organize Information

After you have identified your question, begin gathering and organizing information. There are many ways to gather information, such as researching in a library, interviewing those knowledgeable about the subject, and testing and working in the laboratory and field. Fieldwork is investigations and observations done outside of a laboratory.

Researching Information Before moving in a new direction, it is important to gather the information that already is known about the subject. Start by asking yourself questions to determine exactly what you need to know. Then you will look for the information in various reference sources, like the student is doing in **Figure 1.** Some sources may include textbooks, encyclopedias, government documents, professional journals, science magazines, and the Internet. Always list the sources of your information.

Figure 1 The Internet can be a valuable research tool.

Evaluate Sources of Information Not all sources of information are reliable. You should evaluate all of your sources of information, and use only those you know to be dependable. For example, if you are researching ways to make homes more energy efficient, a site written by the U.S. Department of Energy would be more reliable than a site written by a company that is trying to sell a new type of weatherproofing material. Also, remember that research always is changing. Consult the most current resources available to you. For example, a 1985 resource about saving energy would not reflect the most recent findings.

Sometimes scientists use data that they did not collect themselves, or conclusions drawn by other researchers. This data must be evaluated carefully. Ask questions about how the data were obtained, if the investigation was carried out properly, and if it has been duplicated exactly with the same results. Would you reach the same conclusion from the data? Only when you have confidence in the data can you believe it is true and feel comfortable using it.

SCIENCE SKILL HANDBOOK

MATH SKILL HANDBOOK

FOLDABLES HANDBOOK

REFERENCE HANDBOOK

GLOSSARY/ GLOSARIO

INDEX

Interpret Scientific Illustrations As you research a topic in science, you will see drawings, diagrams, and photographs to help you understand what you read. Some illustrations are included to help you understand an idea that you can't see easily by yourself, like the tiny particles in an atom in **Figure 2.** A drawing helps many people to remember details more easily and provides examples that clarify difficult concepts or give additional information about the topic you are studying. Most illustrations have labels or a caption to identify or to provide more information.

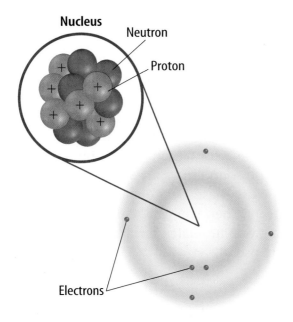

Figure 2 This drawing shows an atom of carbon with its six protons, six neutrons, and six electrons.

Concept Maps One way to organize data is to draw a diagram that shows relationships among ideas (or concepts). A concept map can help make the meanings of ideas and terms more clear, and help you understand and remember what you are studying. Concept maps are useful for breaking large concepts down into smaller parts, making learning easier.

Network Tree A type of concept map that not only shows a relationship, but how the concepts are related is a network tree, shown in **Figure 3.** In a network tree, the words are written in the ovals, while the description of the type of relationship is written across the connecting lines.

When constructing a network tree, write down the topic and all major topics on separate pieces of paper or notecards. Then arrange them in order from general to specific. Branch the related concepts from the major concept and describe the relationship on the connecting line. Continue to more specific concepts until finished.

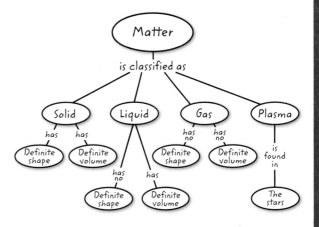

Figure 3 A network tree shows how concepts or objects are related.

Events Chain Another type of concept map is an events chain. Sometimes called a flow chart, it models the order or sequence of items. An events chain can be used to describe a sequence of events, the steps in a procedure, or the stages of a process.

When making an events chain, first find the one event that starts the chain. This event is called the initiating event. Then, find the next event and continue until the outcome is reached, as shown in **Figure 4** on the next page.

SCIENCE SKILL HANDBOOK

MATH SKILL HANDBOOK

FOLDABLES HANDBOOK

REFERENCE HANDBOOK

GLOSSARY/ GLOSARIO

INDEX

SCIENCE SKILL HANDBOOK

MATH SKILL HANDBOOK

FOLDABLES HANDBOOK

REFERENCE HANDBOOK

GLOSSARY/ GLOSARIO

INDEX

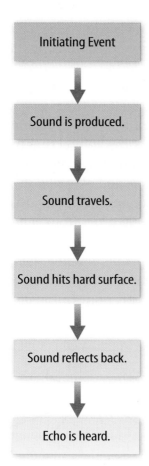

Figure 4 Events-chain concept maps show the order of steps in a process or event. This concept map shows how a sound makes an echo.

Cycle Map A specific type of events chain is a cycle map. It is used when the series of events do not produce a final outcome, but instead relate back to the beginning event, such as in **Figure 5.** Therefore, the cycle repeats itself.

To make a cycle map, first decide what event is the beginning event. This is also called the initiating event. Then list the next events in the order that they occur, with the last event relating back to the initiating event. Words can be written between the events that describe what happens from one event to the next. The number of events in a cycle map can vary, but usually contain three or more events.

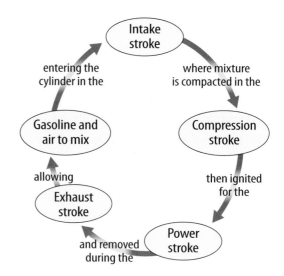

Figure 5 A cycle map shows events that occur in a cycle.

Spider Map A type of concept map that you can use for brainstorming is the spider map. When you have a central idea, you might find that you have a jumble of ideas that relate to it but are not necessarily clearly related to each other. The spider map on sound in **Figure 6** shows that if you write these ideas outside the main concept, then you can begin to separate and group unrelated terms so they become more useful.

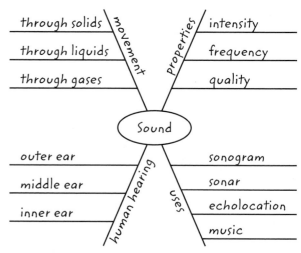

Figure 6 A spider map allows you to list ideas that relate to a central topic but not necessarily to one another.

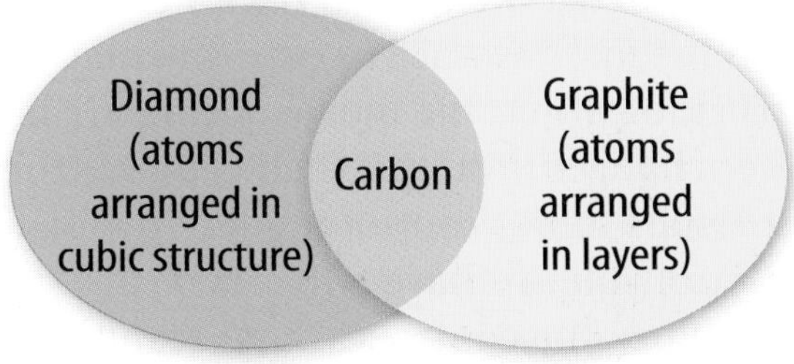

Figure 7 This Venn diagram compares and contrasts two substances made from carbon.

Venn Diagram To illustrate how two subjects compare and contrast you can use a Venn diagram. You can see the characteristics that the subjects have in common and those that they do not, shown in **Figure 7.**

To create a Venn diagram, draw two overlapping ovals that are big enough to write in. List the characteristics unique to one subject in one oval, and the characteristics of the other subject in the other oval. The characteristics in common are listed in the overlapping section.

Make and Use Tables One way to organize information so it is easier to understand is to use a table. Tables can contain numbers, words, or both.

To make a table, list the items to be compared in the first column and the characteristics to be compared in the first row. The title should clearly indicate the content of the table, and the column or row heads should be clear. Notice that in **Table 1** the units are included.

Table 1 Recyclables Collected During Week			
Day of Week	**Paper (kg)**	**Aluminum (kg)**	**Glass (kg)**
Monday	5.0	4.0	12.0
Wednesday	4.0	1.0	10.0
Friday	2.5	2.0	10.0

Make a Model One way to help you better understand the parts of a structure, the way a process works, or to show things too large or small for viewing is to make a model. For example, an atomic model made of a plastic-ball nucleus and chenille stem electron shells can help you visualize how the parts of an atom relate to each other. Other types of models can be devised on a computer or represented by equations.

Form a Hypothesis

A possible explanation based on previous knowledge and observations is called a hypothesis. After researching gasoline types and recalling previous experiences in your family's car you form a hypothesis—our car runs more efficiently because we use premium gasoline. To be valid, a hypothesis has to be something you can test by using an investigation.

Predict When you apply a hypothesis to a specific situation, you predict something about that situation. A prediction makes a statement in advance, based on prior observation, experience, or scientific reasoning. People use predictions to make everyday decisions. Scientists test predictions by performing investigations. Based on previous observations and experiences, you might form a prediction that cars are more efficient with premium gasoline. The prediction can be tested in an investigation.

Design an Experiment A scientist needs to make many decisions before beginning an investigation. Some of these include: how to carry out the investigation, what steps to follow, how to record the data, and how the investigation will answer the question. It also is important to address any safety concerns.

SCIENCE SKILL HANDBOOK

MATH SKILL HANDBOOK

FOLDABLES HANDBOOK

REFERENCE HANDBOOK

GLOSSARY/ GLOSARIO

INDEX

SCIENCE SKILL HANDBOOK

MATH SKILL HANDBOOK

FOLDABLES HANDBOOK

REFERENCE HANDBOOK

GLOSSARY/ GLOSARIO

INDEX

Test the Hypothesis

Now that you have formed your hypothesis, you need to test it. Using an investigation, you will make observations and collect data, or information. This data might either support or not support your hypothesis. Scientists collect and organize data as numbers and descriptions.

Follow a Procedure In order to know what materials to use, as well as how and in what order to use them, you must follow a procedure. **Figure 8** shows a procedure you might follow to test your hypothesis.

Procedure

Step 1	Use regular gasoline for two weeks.
Step 2	Record the number of kilometers between fill-ups and the amount of gasoline used.
Step 3	Switch to premium gasoline for two weeks.
Step 4	Record the number of kilometers between fill-ups and the amount of gasoline used.

Figure 8 A procedure tells you what to do step-by-step.

Identify and Manipulate Variables and Controls In any experiment, it is important to keep everything the same except for the item you are testing. The one factor you change is called the independent variable. The change that results is the dependent variable. Make sure you have only one independent variable, to assure yourself of the cause of the changes you observe in the dependent variable. For example, in your gasoline experiment the type of fuel is the independent variable. The dependent variable is the efficiency.

Many experiments also have a control—an individual instance or experimental subject for which the independent variable is not changed. You can then compare the test results to the control results. To design a control you can have two cars of the same type. The control car uses regular gasoline for four weeks. After you are done with the test, you can compare the experimental results to the control results.

Collect Data

Whether you are carrying out an investigation or a short observational experiment, you will collect data, as shown in **Figure 9.** Scientists collect data as numbers and descriptions and organize them in specific ways.

Observe Scientists observe items and events, then record what they see. When they use only words to describe an observation, it is called qualitative data. Scientists' observations also can describe how much there is of something. These observations use numbers, as well as words, in the description and are called quantitative data. For example, if a sample of the element gold is described as being "shiny and very dense" the data are qualitative. Quantitative data on this sample of gold might include "a mass of 30 g and a density of 19.3 g/cm^3."

Figure 9 Collecting data is one way to gather information directly.

Figure 10 Record data neatly and clearly so it is easy to understand.

When you make observations you should examine the entire object or situation first, and then look carefully for details. It is important to record observations accurately and completely. Always record your notes immediately as you make them, so you do not miss details or make a mistake when recording results from memory. Never put unidentified observations on scraps of paper. Instead they should be recorded in a notebook, like the one in **Figure 10.** Write your data neatly so you can easily read it later. At each point in the experiment, record your observations and label them. That way, you will not have to determine what the figures mean when you look at your notes later. Set up any tables that you will need to use ahead of time, so you can record any observations right away. Remember to avoid bias when collecting data by not including personal thoughts when you record observations. Record only what you observe.

Estimate Scientific work also involves estimating. To estimate is to make a judgment about the size or the number of something without measuring or counting. This is important when the number or size of an object or population is too large or too difficult to accurately count or measure.

Sample Scientists may use a sample or a portion of the total number as a type of estimation. To sample is to take a small, representative portion of the objects or organisms of a population for research. By making careful observations or manipulating variables within that portion of the group, information is discovered and conclusions are drawn that might apply to the whole population. A poorly chosen sample can be unrepresentative of the whole. If you were trying to determine the rainfall in an area, it would not be best to take a rainfall sample from under a tree.

Measure You use measurements every day. Scientists also take measurements when collecting data. When taking measurements, it is important to know how to use measuring tools properly. Accuracy also is important.

Length To measure length, the distance between two points, scientists use meters. Smaller measurements might be measured in centimeters or millimeters.

Length is measured using a metric ruler or meterstick. When using a metric ruler, line up the 0-cm mark with the end of the object being measured and read the number of the unit where the object ends. Look at the metric ruler shown in **Figure 11.** The centimeter lines are the long, numbered lines, and the shorter lines are millimeter lines. In this instance, the length would be 4.50 cm.

Figure 11 This metric ruler has centimeter and millimeter divisions.

SCIENCE SKILL HANDBOOK

MATH SKILL HANDBOOK

FOLDABLES HANDBOOK

REFERENCE HANDBOOK

GLOSSARY/ GLOSARIO

INDEX

SCIENCE SKILL HANDBOOK

MATH SKILL HANDBOOK

FOLDABLES HANDBOOK

REFERENCE HANDBOOK

GLOSSARY/ GLOSARIO

INDEX

Mass The SI unit for mass is the kilogram (kg). Scientists can measure mass using units formed by adding metric prefixes to the unit gram (g), such as milligram (mg). To measure mass, you might use a triple-beam balance similar to the one shown in **Figure 12.** The balance has a pan on one side and a set of beams on the other side. Each beam has a rider that slides on the beam.

When using a triple-beam balance, place an object on the pan. Slide the largest rider along its beam until the pointer drops below zero. Then move it back one notch. Repeat the process for each rider proceeding from the larger to smaller until the pointer swings an equal distance above and below the zero point. Sum the masses on each beam to find the mass of the object. Move all riders back to zero when finished.

Instead of putting materials directly on the balance, scientists often take a tare of a container. A tare is the mass of a container into which objects or substances are placed for measuring their masses. To find the mass of objects or substances, find the mass of a clean container. Remove the container from the pan, and place the object or substances in the container. Find the mass of the container with the materials in it. Subtract the mass of the empty container from the mass of the filled container to find the mass of the materials you are using.

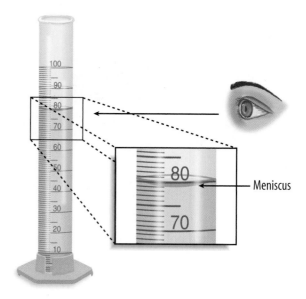

Figure 13 Graduated cylinders measure liquid volume.

Liquid Volume To measure liquids, the unit used is the liter. When a smaller unit is needed, scientists might use a milliliter. Because a milliliter takes up the volume of a cube measuring 1 cm on each side it also can be called a cubic centimeter ($cm^3 = cm \times cm \times cm$).

You can use beakers and graduated cylinders to measure liquid volume. A graduated cylinder, shown in **Figure 13,** is marked from bottom to top in milliliters. In lab, you might use a 10-mL graduated cylinder or a 100-mL graduated cylinder. When measuring liquids, notice that the liquid has a curved surface. Look at the surface at eye level, and measure the bottom of the curve. This is called the meniscus. The graduated cylinder in **Figure 13** contains 79.0 mL, or 79.0 cm^3, of a liquid.

Temperature Scientists often measure temperature using the Celsius scale. Pure water has a freezing point of 0°C and boiling point of 100°C. The unit of measurement is degrees Celsius. Two other scales often used are the Fahrenheit and Kelvin scales.

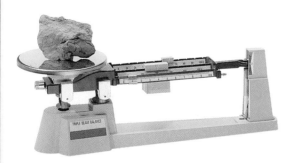

Figure 12 A triple-beam balance is used to determine the mass of an object.

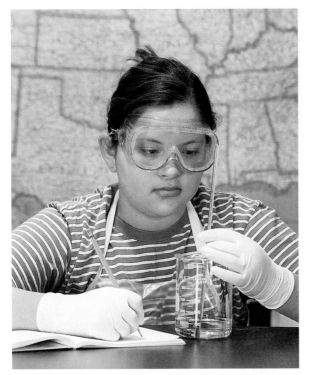

Figure 14 A thermometer measures the temperature of an object.

Scientists use a thermometer to measure temperature. Most thermometers in a laboratory are glass tubes with a bulb at the bottom end containing a liquid such as colored alcohol. The liquid rises or falls with a change in temperature. To read a glass thermometer like the thermometer in **Figure 14,** rotate it slowly until a red line appears. Read the temperature where the red line ends.

Form Operational Definitions An operational definition defines an object by how it functions, works, or behaves. For example, when you are playing hide and seek and a tree is home base, you have created an operational definition for a tree.

Objects can have more than one operational definition. For example, a ruler can be defined as a tool that measures the length of an object (how it is used). It can also be a tool with a series of marks used as a standard when measuring (how it works).

Analyze the Data

To determine the meaning of your observations and investigation results, you will need to look for patterns in the data. Then you must think critically to determine what the data mean. Scientists use several approaches when they analyze the data they have collected and recorded. Each approach is useful for identifying specific patterns.

Interpret Data The word *interpret* means "to explain the meaning of something." When analyzing data from an experiment, try to find out what the data show. Identify the control group and the test group to see whether changes in the independent variable have had an effect. Look for differences in the dependent variable between the control and test groups.

Classify Sorting objects or events into groups based on common features is called classifying. When classifying, first observe the objects or events to be classified. Then select one feature that is shared by some members in the group, but not by all. Place those members that share that feature in a subgroup. You can classify members into smaller and smaller subgroups based on characteristics. Remember that when you classify, you are grouping objects or events for a purpose. Keep your purpose in mind as you select the features to form groups and subgroups.

Compare and Contrast Observations can be analyzed by noting the similarities and differences between two or more objects or events that you observe. When you look at objects or events to see how they are similar, you are comparing them. Contrasting is looking for differences in objects or events.

SCIENCE SKILL HANDBOOK

MATH SKILL HANDBOOK

FOLDABLES HANDBOOK

REFERENCE HANDBOOK

GLOSSARY/ GLOSARIO

INDEX

SCIENCE SKILL HANDBOOK

MATH SKILL HANDBOOK

FOLDABLES HANDBOOK

REFERENCE HANDBOOK

GLOSSARY/ GLOSARIO

INDEX

Recognize Cause and Effect A cause is a reason for an action or condition. The effect is that action or condition. When two events happen together, it is not necessarily true that one event caused the other. Scientists must design a controlled investigation to recognize the exact cause and effect.

Draw Conclusions

When scientists have analyzed the data they collected, they proceed to draw conclusions about the data. These conclusions are sometimes stated in words similar to the hypothesis that you formed earlier. They may confirm a hypothesis, or lead you to a new hypothesis.

Infer Scientists often make inferences based on their observations. An inference is an attempt to explain observations or to indicate a cause. An inference is not a fact, but a logical conclusion that needs further investigation. For example, you may infer that a fire has caused smoke. Until you investigate, however, you do not know for sure.

Apply When you draw a conclusion, you must apply those conclusions to determine whether the data supports the hypothesis. If your data do not support your hypothesis, it does not mean that the hypothesis is wrong. It means only that the result of the investigation did not support the hypothesis. Maybe the experiment needs to be redesigned, or some of the initial observations on which the hypothesis was based were incomplete or biased. Perhaps more observation or research is needed to refine your hypothesis. A successful investigation does not always come out the way you originally predicted.

Avoid Bias Sometimes a scientific investigation involves making judgments. When you make a judgment, you form an opinion. It is important to be honest and not to allow any expectations of results to bias your judgments. This is important throughout the entire investigation, from researching to collecting data to drawing conclusions.

Communicate

The communication of ideas is an important part of the work of scientists. A discovery that is not reported will not advance the scientific community's understanding or knowledge. Communication among scientists also is important as a way of improving their investigations.

Scientists communicate in many ways, from writing articles in journals and magazines that explain their investigations and experiments, to announcing important discoveries on television and radio. Scientists also share ideas with colleagues on the Internet or present them as lectures, like the student is doing in **Figure 15.**

Figure 15 A student communicates to his peers about his investigation.

These safety symbols are used in laboratory and field investigations in this book to indicate possible hazards. Learn the meaning of each symbol and refer to this page often. *Remember to wash your hands thoroughly after completing lab procedures.*

PROTECTIVE EQUIPMENT   Do not begin any lab without the proper protection equipment.

GOGGLES	Proper eye protection must be worn when performing or observing science activities that involve items or conditions as listed below.	**APRON**	Wear an approved apron when using substances that could stain, wet, or destroy cloth.	**SOAP**	Wash hands with soap and water before removing goggles and after all lab activities.	**GLOVES**	Wear gloves when working with biological materials, chemicals, animals, or materials that can stain or irritate hands.

LABORATORY HAZARDS

Symbols	Potential Hazards	Precaution	Response
DISPOSAL	contamination of classroom or environment due to improper disposal of materials such as chemicals and live specimens	• DO NOT dispose of hazardous materials in the sink or trash can. • Dispose of wastes as directed by your teacher.	• If hazardous materials are disposed of improperly, notify your teacher immediately.
EXTREME TEMPERATURE	skin burns due to extremely hot or cold materials such as hot glass, liquids, or metals; liquid nitrogen; dry ice	• Use proper protective equipment, such as hot mitts and/or tongs, when handling objects with extreme temperatures.	• If injury occurs, notify your teacher immediately.
SHARP OBJECTS	punctures or cuts from sharp objects such as razor blades, pins, scalpels, and broken glass	• Handle glassware carefully to avoid breakage. • Walk with sharp objects pointed downward, away from you and others.	• If broken glass or injury occurs, notify your teacher immediately.
ELECTRICAL	electric shock or skin burn due to improper grounding, short circuits, liquid spills, or exposed wires	• Check condition of wires and apparatus for fraying or uninsulated wires, and broken or cracked equipment. • Use only GFCI-protected outlets	• DO NOT attempt to fix electrical problems. Notify your teacher immediately.
CHEMICAL	skin irritation or burns, breathing difficulty, and/or poisoning due to touching, swallowing, or inhalation of chemicals such as acids, bases, bleach, metal compounds, iodine, poinsettias, pollen, ammonia, acetone, nail polish remover, heated chemicals, mothballs, and any other chemicals labeled or known to be dangerous	• Wear proper protective equipment such as goggles, apron, and gloves when using chemicals. • Ensure proper room ventilation or use a fume hood when using materials that produce fumes. • NEVER smell fumes directly. • NEVER taste or eat any material in the laboratory.	• If contact occurs, immediately flush affected area with water and notify your teacher. • If a spill occurs, leave the area immediately and notify your teacher.
FLAMMABLE	unexpected fire due to liquids or gases that ignite easily such as rubbing alcohol	• Avoid open flames, sparks, or heat when flammable liquids are present.	• If a fire occurs, leave the area immediately and notify your teacher.
OPEN FLAME	burns or fire due to open flame from matches, Bunsen burners, or burning materials	• Tie back loose hair and clothing. • Keep flame away from all materials. • Follow teacher instructions when lighting and extinguishing flames. • Use proper protection, such as hot mitts or tongs, when handling hot objects.	• If a fire occurs, leave the area immediately and notify your teacher.
ANIMAL SAFETY	injury to or from laboratory animals	• Wear proper protective equipment such as gloves, apron, and goggles when working with animals. • Wash hands after handling animals.	• If injury occurs, notify your teacher immediately.
BIOLOGICAL	infection or adverse reaction due to contact with organisms such as bacteria, fungi, and biological materials such as blood, animal or plant materials	• Wear proper protective equipment such as gloves, goggles, and apron when working with biological materials. • Avoid skin contact with an organism or any part of the organism. • Wash hands after handling organisms.	• If contact occurs, wash the affected area and notify your teacher immediately.
FUME	breathing difficulties from inhalation of fumes from substances such as ammonia, acetone, nail polish remover, heated chemicals, and mothballs	• Wear goggles, apron, and gloves. • Ensure proper room ventilation or use a fume hood when using substances that produce fumes. • NEVER smell fumes directly.	• If a spill occurs, leave area and notify your teacher immediately.
IRRITANT	irritation of skin, mucous membranes, or respiratory tract due to materials such as acids, bases, bleach, pollen, mothballs, steel wool, and potassium permanganate	• Wear goggles, apron, and gloves. • Wear a dust mask to protect against fine particles.	• If skin contact occurs, immediately flush the affected area with water and notify your teacher.
RADIOACTIVE	excessive exposure from alpha, beta, and gamma particles	• Remove gloves and wash hands with soap and water before removing remainder of protective equipment.	• If cracks or holes are found in the container, notify your teacher immediately.

SCIENCE SKILL HANDBOOK

MATH SKILL HANDBOOK

FOLDABLES HANDBOOK

REFERENCE HANDBOOK

GLOSSARY/ GLOSARIO

INDEX

Safety in the Science Laboratory

Introduction to Science Safety

The science laboratory is a safe place to work if you follow standard safety procedures. Being responsible for your own safety helps to make the entire laboratory a safer place for everyone. When performing any lab, read and apply the caution statements and safety symbol listed at the beginning of the lab.

General Safety Rules

1. Complete the *Lab Safety Form* or other safety contract BEFORE starting any science lab.

2. Study the procedure. Ask your teacher any questions. Be sure you understand safety symbols shown on the page.

3. Notify your teacher about allergies or other health conditions that can affect your participation in a lab.

4. Learn and follow use and safety procedures for your equipment. If unsure, ask your teacher.

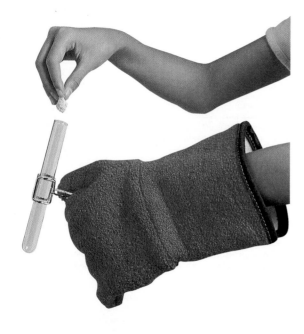

5. Never eat, drink, chew gum, apply cosmetics, or do any personal grooming in the lab. Never use lab glassware as food or drink containers. Keep your hands away from your face and mouth.

6. Know the location and proper use of the safety shower, eye wash, fire blanket, and fire alarm.

Prevent Accidents

1. Use the safety equipment provided to you. Goggles and a safety apron should be worn during investigations.

2. Do NOT use hair spray, mousse, or other flammable hair products. Tie back long hair and tie down loose clothing.

3. Do NOT wear sandals or other open-toed shoes in the lab.

4. Remove jewelry on hands and wrists. Loose jewelry, such as chains and long necklaces, should be removed to prevent them from getting caught in equipment.

5. Do not taste any substances or draw any material into a tube with your mouth.

6. Proper behavior is expected in the lab. Practical jokes and fooling around can lead to accidents and injury.

7. Keep your work area uncluttered.

Laboratory Work

1. Collect and carry all equipment and materials to your work area before beginning a lab.

2. Remain in your own work area unless given permission by your teacher to leave it.

SCIENCE SKILL HANDBOOK

MATH SKILL HANDBOOK

FOLDABLES HANDBOOK

REFERENCE HANDBOOK

GLOSSARY/ GLOSARIO

INDEX

3. Always slant test tubes away from your-self and others when heating them, adding substances to them, or rinsing them.

4. If instructed to smell a substance in a container, hold the container a short distance away and fan vapors toward your nose.

5. Do NOT substitute other chemicals/substances for those in the materials list unless instructed to do so by your teacher.

6. Do NOT take any materials or chemicals outside of the laboratory.

7. Stay out of storage areas unless instructed to be there and supervised by your teacher.

Laboratory Cleanup

1. Turn off all burners, water, and gas, and disconnect all electrical devices.

2. Clean all pieces of equipment and return all materials to their proper places.

3. Dispose of chemicals and other materials as directed by your teacher. Place broken glass and solid substances in the proper containers. Never discard materials in the sink.

4. Clean your work area.

5. Wash your hands with soap and water thoroughly BEFORE removing your goggles.

Emergencies

1. Report any fire, electrical shock, glass-ware breakage, spill, or injury, no matter how small, to your teacher immediately. Follow his or her instructions.

2. If your clothing should catch fire, STOP, DROP, and ROLL. If possible, smother it with the fire blanket or get under a safety shower. NEVER RUN.

3. If a fire should occur, turn off all gas and leave the room according to established procedures.

4. In most instances, your teacher will clean up spills. Do NOT attempt to clean up spills unless you are given permission and instructions to do so.

5. If chemicals come into contact with your eyes or skin, notify your teacher immediately. Use the eyewash, or flush your skin or eyes with large quantities of water.

6. The fire extinguisher and first-aid kit should only be used by your teacher unless it is an extreme emergency and you have been given permission.

7. If someone is injured or becomes ill, only a professional medical provider or someone certified in first aid should perform first-aid procedures.

SCIENCE SKILL HANDBOOK

MATH SKILL HANDBOOK

FOLDABLES HANDBOOK

REFERENCE HANDBOOK

GLOSSARY/ GLOSARIO

INDEX

Math Review

SCIENCE SKILL HANDBOOK

MATH SKILL HANDBOOK

FOLDABLES HANDBOOK

REFERENCE HANDBOOK

GLOSSARY/ GLOSARIO

INDEX

Use Fractions

A fraction compares a part to a whole. In the fraction $\frac{2}{3}$, the 2 represents the part and is the numerator. The 3 represents the whole and is the denominator.

Reduce Fractions To reduce a fraction, you must find the largest factor that is common to both the numerator and the denominator, the greatest common factor (GCF). Divide both numbers by the GCF. The fraction has then been reduced, or it is in its simplest form.

Example

Twelve of the 20 chemicals in the science lab are in powder form. What fraction of the chemicals used in the lab are in powder form?

Step 1 Write the fraction.

$$\frac{part}{whole} = \frac{12}{20}$$

Step 2 To find the GCF of the numerator and denominator, list all of the factors of each number.

Factors of 12: 1, 2, 3, 4, 6, 12 (the numbers that divide evenly into 12)

Factors of 20: 1, 2, 4, 5, 10, 20 (the numbers that divide evenly into 20)

Step 3 List the common factors.

1, 2, 4

Step 4 Choose the greatest factor in the list. The GCF of 12 and 20 is 4.

Step 5 Divide the numerator and denominator by the GCF.

$$\frac{12 \div 4}{20 \div 4} = \frac{3}{5}$$

In the lab, $\frac{3}{5}$ of the chemicals are in powder form.

Practice Problem At an amusement park, 66 of 90 rides have a height restriction. What fraction of the rides, in its simplest form, has a height restriction?

Add and Subtract Fractions with Like Denominators To add or subtract fractions with the same denominator, add or subtract the numerators and write the sum or difference over the denominator. After finding the sum or difference, find the simplest form for your fraction.

Example 1

In the forest outside your house, $\frac{1}{8}$ of the animals are rabbits, $\frac{3}{8}$ are squirrels, and the remainder are birds and insects. How many are mammals?

Step 1 Add the numerators.

$$\frac{1}{8} + \frac{3}{8} = \frac{(1 + 3)}{8} = \frac{4}{8}$$

Step 2 Find the GCF.

$$\frac{4}{8} \text{ (GCF, 4)}$$

Step 3 Divide the numerator and denominator by the GCF.

$$\frac{4 \div 4}{8 \div 4} = \frac{1}{2}$$

$\frac{1}{2}$ of the animals are mammals.

Example 2

If $\frac{7}{16}$ of the Earth is covered by freshwater, and $\frac{1}{16}$ of that is in glaciers, how much freshwater is not frozen?

Step 1 Subtract the numerators.

$$\frac{7}{16} - \frac{1}{16} = \frac{(7 - 1)}{16} = \frac{6}{16}$$

Step 2 Find the GCF.

$$\frac{6}{16} \text{ (GCF, 2)}$$

Step 3 Divide the numerator and denominator by the GCF.

$$\frac{6 \div 2}{16 \div 2} = \frac{3}{8}$$

$\frac{3}{8}$ of the freshwater is not frozen.

Practice Problem A bicycle rider is riding at a rate of 15 km/h for $\frac{4}{9}$ of his ride, 10 km/h for $\frac{2}{9}$ of his ride, and 8 km/h for the remainder of the ride. How much of his ride is he riding at a rate greater than 8 km/h?

Add and Subtract Fractions with Unlike Denominators To add or subtract fractions with unlike denominators, first find the least common denominator (LCD). This is the smallest number that is a common multiple of both denominators. Rename each fraction with the LCD, and then add or subtract. Find the simplest form if necessary.

Example 1

A chemist makes a paste that is $\frac{1}{2}$ table salt (NaCl), $\frac{1}{3}$ sugar ($C_6H_{12}O_6$), and the remainder is water (H_2O). How much of the paste is a solid?

Step 1 Find the LCD of the fractions.

$$\frac{1}{2} + \frac{1}{3} \text{ (LCD, 6)}$$

Step 2 Rename each numerator and each denominator with the LCD.

Step 3 Add the numerators.

$$\frac{3}{6} + \frac{2}{6} = \frac{(3+2)}{6} = \frac{5}{6}$$

$\frac{5}{6}$ of the paste is a solid.

Example 2

The average precipitation in Grand Junction, CO, is $\frac{7}{10}$ inch in November, and $\frac{3}{5}$ inch in December. What is the total average precipitation?

Step 1 Find the LCD of the fractions.

$$\frac{7}{10} + \frac{3}{5} \text{ (LCD, 10)}$$

Step 2 Rename each numerator and each denominator with the LCD.

Step 3 Add the numerators.

$$\frac{7}{10} + \frac{6}{10} = \frac{(7+6)}{10} = \frac{13}{10}$$

$\frac{13}{10}$ inches total precipitation, or $1\frac{3}{10}$ inches.

Practice Problem On an electric bill, about $\frac{1}{8}$ of the energy is from solar energy and about $\frac{1}{10}$ is from wind power. How much of the total bill is from solar energy and wind power combined?

Example 3

In your body, $\frac{7}{10}$ of your muscle contractions are involuntary (cardiac and smooth muscle tissue). Smooth muscle makes $\frac{3}{15}$ of your muscle contractions. How many of your muscle contractions are made by cardiac muscle?

Step 1 Find the LCD of the fractions.

$$\frac{7}{10} - \frac{3}{15} \text{ (LCD, 30)}$$

Step 2 Rename each numerator and each denominator with the LCD.

$$\frac{7 \times 3}{10 \times 3} = \frac{21}{30}$$

$$\frac{3 \times 2}{15 \times 2} = \frac{6}{30}$$

Step 3 Subtract the numerators.

$$\frac{21}{30} - \frac{6}{30} = \frac{(21-6)}{30} = \frac{15}{30}$$

Step 4 Find the GCF.

$$\frac{15}{30} \text{ (GCF, 15)}$$

$$\frac{1}{2}$$

$\frac{1}{2}$ of all muscle contractions are cardiac muscle.

Example 4

Tony wants to make cookies that call for $\frac{3}{4}$ of a cup of flour, but he only has $\frac{1}{3}$ of a cup. How much more flour does he need?

Step 1 Find the LCD of the fractions.

$$\frac{3}{4} - \frac{1}{3} \text{ (LCD, 12)}$$

Step 2 Rename each numerator and each denominator with the LCD.

$$\frac{3 \times 3}{4 \times 3} = \frac{9}{12}$$

$$\frac{1 \times 4}{3 \times 4} = \frac{4}{12}$$

Step 3 Subtract the numerators.

$$\frac{9}{12} - \frac{4}{12} = \frac{(9-4)}{12} = \frac{5}{12}$$

$\frac{5}{12}$ of a cup of flour

Practice Problem Using the information provided to you in Example 3 above, determine how many muscle contractions are voluntary (skeletal muscle).

SCIENCE SKILL HANDBOOK

MATH SKILL HANDBOOK

FOLDABLES HANDBOOK

REFERENCE HANDBOOK

GLOSSARY/ GLOSARIO

INDEX

Multiply Fractions To multiply with fractions, multiply the numerators and multiply the denominators. Find the simplest form if necessary.

> **Example**
>
> Multiply $\frac{3}{5}$ by $\frac{1}{3}$.
>
> **Step 1** Multiply the numerators and denominators.
>
> $$\frac{3}{5} \times \frac{1}{3} = \frac{(3 \times 1)}{(5 \times 3)} \frac{3}{15}$$
>
> **Step 2** Find the GCF.
>
> $$\frac{3}{15} \text{ (GCF, 3)}$$
>
> **Step 3** Divide the numerator and denominator by the GCF.
>
> $$\frac{3 \div 3}{15 \div 3} = \frac{1}{5}$$
>
> $\frac{3}{5}$ multiplied by $\frac{1}{3}$ is $\frac{1}{5}$.

Practice Problem Multiply $\frac{3}{14}$ by $\frac{5}{16}$.

Find a Reciprocal Two numbers whose product is 1 are called multiplicative inverses, or reciprocals.

> **Example**
>
> Find the reciprocal of $\frac{3}{8}$.
>
> **Step 1** Inverse the fraction by putting the denominator on top and the numerator on the bottom.
>
> $$\frac{8}{3}$$
>
> The reciprocal of $\frac{3}{8}$ is $\frac{8}{3}$.

Practice Problem Find the reciprocal of $\frac{4}{9}$.

Divide Fractions To divide one fraction by another fraction, multiply the dividend by the reciprocal of the divisor. Find the simplest form if necessary.

> **Example 1**
>
> Divide $\frac{1}{9}$ by $\frac{1}{3}$.
>
> **Step 1** Find the reciprocal of the divisor.
>
> The reciprocal of $\frac{1}{3}$ is $\frac{3}{1}$.
>
> **Step 2** Multiply the dividend by the reciprocal of the divisor.
>
> $$\frac{\frac{1}{9}}{\frac{1}{3}} = \frac{1}{9} \times \frac{3}{1} = \frac{(1 \times 3)}{(9 \times 1)} = \frac{3}{9}$$
>
> **Step 3** Find the GCF.
>
> $$\frac{3}{9} \text{ (GCF, 3)}$$
>
> **Step 4** Divide the numerator and denominator by the GCF.
>
> $$\frac{3 \div 3}{9 \div 3} = \frac{1}{3}$$
>
> $\frac{1}{9}$ divided by $\frac{1}{3}$ is $\frac{1}{3}$.

> **Example 2**
>
> Divide $\frac{3}{5}$ by $\frac{1}{4}$.
>
> **Step 1** Find the reciprocal of the divisor.
>
> The reciprocal of $\frac{1}{4}$ is $\frac{4}{1}$.
>
> **Step 2** Multiply the dividend by the reciprocal of the divisor.
>
> $$\frac{\frac{3}{5}}{\frac{1}{4}} = \frac{3}{5} \times \frac{4}{1} = \frac{(3 \times 4)}{(5 \times 1)} = \frac{12}{5}$$
>
> $\frac{3}{5}$ divided by $\frac{1}{4}$ is $\frac{12}{5}$ or $2\frac{2}{5}$.

Practice Problem Divide $\frac{3}{11}$ by $\frac{7}{10}$.

SCIENCE SKILL HANDBOOK

MATH SKILL HANDBOOK

FOLDABLES HANDBOOK

REFERENCE HANDBOOK

GLOSSARY/ GLOSARIO

INDEX

Use Ratios

When you compare two numbers by division, you are using a ratio. Ratios can be written 3 to 5, 3:5, or $\frac{3}{5}$. Ratios, like fractions, also can be written in simplest form.

Ratios can represent one type of probability, called odds. This is a ratio that compares the number of ways a certain outcome occurs to the number of possible outcomes. For example, if you flip a coin 100 times, what are the odds that it will come up heads? There are two possible outcomes, heads or tails, so the odds of coming up heads are 50:100. Another way to say this is that 50 out of 100 times the coin will come up heads. In its simplest form, the ratio is 1:2.

Example 1

A chemical solution contains 40 g of salt and 64 g of baking soda. What is the ratio of salt to baking soda as a fraction in simplest form?

Step 1   Write the ratio as a fraction.

$$\frac{salt}{baking\ soda} = \frac{40}{64}$$

Step 2   Express the fraction in simplest form. The GCF of 40 and 64 is 8.

$$\frac{40}{64} = \frac{40 \div 8}{64 \div 8} = \frac{5}{8}$$

The ratio of salt to baking soda in the sample is 5:8.

Example 2

Sean rolls a 6-sided die 6 times. What are the odds that the side with a 3 will show?

Step 1   Write the ratio as a fraction.

$$\frac{number\ of\ sides\ with\ a\ 3}{number\ of\ possible\ sides} = \frac{1}{6}$$

Step 2   Multiply by the number of attempts.

$$\frac{1}{6} \times 6\ attempts = \frac{6}{6}\ attempts = 1\ attempt$$

1 attempt out of 6 will show a 3.

Practice Problem Two metal rods measure 100 cm and 144 cm in length. What is the ratio of their lengths in simplest form?

Use Decimals

A fraction with a denominator that is a power of ten can be written as a decimal. For example, 0.27 means $\frac{27}{100}$. The decimal point separates the ones place from the tenths place.

Any fraction can be written as a decimal using division. For example, the fraction $\frac{5}{8}$ can be written as a decimal by dividing 5 by 8. Written as a decimal, it is 0.625.

Add or Subtract Decimals When adding and subtracting decimals, line up the decimal points before carrying out the operation.

Example 1

Find the sum of 47.68 and 7.80.

Step 1   Line up the decimal places when you write the numbers.

$$\begin{array}{r} 47.68 \\ + 7.80 \\ \hline \end{array}$$

Step 2   Add the decimals.

$$\begin{array}{r} \overset{1\ 1}{47.68} \\ + 7.80 \\ \hline 55.48 \end{array}$$

The sum of 47.68 and 7.80 is 55.48.

Example 2

Find the difference of 42.17 and 15.85.

Step 1   Line up the decimal places when you write the number.

$$\begin{array}{r} 42.17 \\ -15.85 \\ \hline \end{array}$$

Step 2   Subtract the decimals.

$$\begin{array}{r} \overset{3\ 11}{4\cancel{2}.17} \\ -15.85 \\ \hline 26.32 \end{array}$$

The difference of 42.17 and 15.85 is 26.32.

Practice Problem Find the sum of 1.245 and 3.842.

SCIENCE SKILL HANDBOOK

MATH SKILL HANDBOOK

FOLDABLES HANDBOOK

REFERENCE HANDBOOK

GLOSSARY/ GLOSARIO

INDEX

Multiply Decimals To multiply decimals, multiply the numbers like numbers without decimal points. Count the decimal places in each factor. The product will have the same number of decimal places as the sum of the decimal places in the factors.

Example

Multiply 2.4 by 5.9.

Step 1 Multiply the factors like two whole numbers.

$24 \times 59 = 1416$

Step 2 Find the sum of the number of decimal places in the factors. Each factor has one decimal place, for a sum of two decimal places.

Step 3 The product will have two decimal places.

14.16

The product of 2.4 and 5.9 is 14.16.

Practice Problem Multiply 4.6 by 2.2.

Divide Decimals When dividing decimals, change the divisor to a whole number. To do this, multiply both the divisor and the dividend by the same power of ten. Then place the decimal point in the quotient directly above the decimal point in the dividend. Then divide as you do with whole numbers.

Example

Divide 8.84 by 3.4.

Step 1 Multiply both factors by 10.

$3.4 \times 10 = 34, 8.84 \times 10 = 88.4$

Step 2 Divide 88.4 by 34.

$$\begin{array}{r} 2.6 \\ 34\overline{)88.4} \\ -68 \\ \hline 204 \\ -204 \\ \hline 0 \end{array}$$

8.84 divided by 3.4 is 2.6.

Practice Problem Divide 75.6 by 3.6.

Use Proportions

An equation that shows that two ratios are equivalent is a proportion. The ratios $\frac{2}{4}$ and $\frac{5}{10}$ are equivalent, so they can be written as $\frac{2}{4} = \frac{5}{10}$. This equation is a proportion.

When two ratios form a proportion, the cross products are equal. To find the cross products in the proportion $\frac{2}{4} = \frac{5}{10}$, multiply the 2 and the 10, and the 4 and the 5. Therefore $2 \times 10 = 4 \times 5$, or $20 = 20$.

Because you know that both ratios are equal, you can use cross products to find a missing term in a proportion. This is known as solving the proportion.

Example

The heights of a tree and a pole are proportional to the lengths of their shadows. The tree casts a shadow of 24 m when a 6-m pole casts a shadow of 4 m. What is the height of the tree?

Step 1 Write a proportion.

$$\frac{\text{height of tree}}{\text{height of pole}} = \frac{\text{length of tree's shadow}}{\text{length of pole's shadow}}$$

Step 2 Substitute the known values into the proportion. Let h represent the unknown value, the height of the tree.

$$\frac{h}{6} \times \frac{24}{4}$$

Step 3 Find the cross products.

$h \times 4 = 6 \times 24$

Step 4 Simplify the equation.

$4h \times 144$

Step 5 Divide each side by 4.

$$\frac{4h}{4} \times \frac{144}{4}$$

$h = 36$

The height of the tree is 36 m.

Practice Problem The ratios of the weights of two objects on the Moon and on Earth are in proportion. A rock weighing 3 N on the Moon weighs 18 N on Earth. How much would a rock that weighs 5 N on the Moon weigh on Earth?

SR-18 • Math Skill Handbook

Use Percentages

The word *percent* means "out of one hundred." It is a ratio that compares a number to 100. Suppose you read that 77 percent of Earth's surface is covered by water. That is the same as reading that the fraction of Earth's surface covered by water is $\frac{77}{100}$. To express a fraction as a percent, first find the equivalent decimal for the fraction. Then, multiply the decimal by 100 and add the percent symbol.

Example 1

Express $\frac{13}{20}$ as a percent.

Step 1 Find the equivalent decimal for the fraction.

$$
\begin{array}{r}
0.65 \\
20\overline{)13.00} \\
\underline{12\,0} \\
1\,00 \\
\underline{1\,00} \\
0
\end{array}
$$

Step 2 Rewrite the fraction $\frac{13}{20}$ as 0.65.

Step 3 Multiply 0.65 by 100 and add the % symbol.

$$0.65 \times 100 = 65 = 65\%$$

So, $\frac{13}{20} = 65\%$.

This also can be solved as a proportion.

Example 2

Express $\frac{13}{20}$ as a percent.

Step 1 Write a proportion.

$$\frac{13}{20} = \frac{x}{100}$$

Step 2 Find the cross products.

$$1300 = 20x$$

Step 3 Divide each side by 20.

$$\frac{1300}{20} = \frac{20x}{20}$$

$$65\% = x$$

Practice Problem In one year, 73 of 365 days were rainy in one city. What percent of the days in that city were rainy?

Solve One-Step Equations

A statement that two expressions are equal is an equation. For example, $A = B$ is an equation that states that A is equal to B.

An equation is solved when a variable is replaced with a value that makes both sides of the equation equal. To make both sides equal the inverse operation is used. Addition and subtraction are inverses, and multiplication and division are inverses.

Example 1

Solve the equation $x - 10 = 35$.

Step 1 Find the solution by adding 10 to each side of the equation.

$$x - 10 = 35$$
$$x - 10 + 10 = 35 - 10$$
$$x = 45$$

Step 2 Check the solution.

$$x - 10 = 35$$
$$45 - 10 = 35$$
$$35 = 35$$

Both sides of the equation are equal, so $x = 45$.

Example 2

In the formula $a = bc$, find the value of c if $a = 20$ and $b = 2$.

Step 1 Rearrange the formula so the unknown value is by itself on one side of the equation by dividing both sides by b.

$$a = bc$$
$$\frac{a}{b} = \frac{bc}{b}$$
$$\frac{a}{b} = c$$

Step 2 Replace the variables a and b with the values that are given.

$$\frac{a}{b} = c$$
$$\frac{20}{2} = c$$
$$10 = c$$

Step 3 Check the solution.

$$a = bc$$
$$20 = 2 \times 10$$
$$20 = 20$$

Both sides of the equation are equal, so $c = 10$ is the solution when $a = 20$ and $b = 2$.

Practice Problem In the formula $h = gd$, find the value of d if $g = 12.3$ and $h = 17.4$.

SCIENCE SKILL HANDBOOK

MATH SKILL HANDBOOK

FOLDABLES HANDBOOK

REFERENCE HANDBOOK

GLOSSARY/GLOSARIO

INDEX

Use Statistics

The branch of mathematics that deals with collecting, analyzing, and presenting data is statistics. In statistics, there are three common ways to summarize data with a single number—the mean, the median, and the mode.

The **mean** of a set of data is the arithmetic average. It is found by adding the numbers in the data set and dividing by the number of items in the set.

The **median** is the middle number in a set of data when the data are arranged in numerical order. If there were an even number of data points, the median would be the mean of the two middle numbers.

The **mode** of a set of data is the number or item that appears most often.

Another number that often is used to describe a set of data is the range. The **range** is the difference between the largest number and the smallest number in a set of data.

Example

The speeds (in m/s) for a race car during five different time trials are 39, 37, 44, 36, and 44.

To find the mean:

Step 1 Find the sum of the numbers.

$$39 + 37 + 44 + 36 + 44 = 200$$

Step 2 Divide the sum by the number of items, which is 5.

$$200 \div 5 = 40$$

The mean is 40 m/s.

To find the median:

Step 1 Arrange the measures from least to greatest.

36, 37, 39, 44, 44

Step 2 Determine the middle measure.

36, 37, <u>39</u>, 44, 44

The median is 39 m/s.

To find the mode:

Step 1 Group the numbers that are the same together.

44, 44, 36, 37, 39

Step 2 Determine the number that occurs most in the set.

<u>44, 44,</u> 36, 37, 39

The mode is 44 m/s.

To find the range:

Step 1 Arrange the measures from greatest to least.

44, 44, 39, 37, 36

Step 2 Determine the greatest and least measures in the set.

<u>44,</u> 44, 39, 37, <u>36</u>

Step 3 Find the difference between the greatest and least measures.

$$44 - 36 = 8$$

The range is 8 m/s.

Practice Problem Find the mean, median, mode, and range for the data set 8, 4, 12, 8, 11, 14, 16.

A **frequency table** shows how many times each piece of data occurs, usually in a survey. **Table 1** below shows the results of a student survey on favorite color.

Table 1 Student Color Choice		
Color	**Tally**	**Frequency**
red	IIII	4
blue	IIII	5
black	II	2
green	III	3
purple	IIII II	7
yellow	IIII I	6

Based on the frequency table data, which color is the favorite?

SCIENCE SKILL HANDBOOK

MATH SKILL HANDBOOK

FOLDABLES HANDBOOK

REFERENCE HANDBOOK

GLOSSARY/ GLOSARIO

INDEX

Use Geometry

The branch of mathematics that deals with the measurement, properties, and relationships of points, lines, angles, surfaces, and solids is called geometry.

Perimeter The **perimeter** (P) is the distance around a geometric figure. To find the perimeter of a rectangle, add the length and width and multiply that sum by two, or $2(l + w)$. To find perimeters of irregular figures, add the length of the sides.

Example 1

Find the perimeter of a rectangle that is 3 m long and 5 m wide.

Step 1   You know that the perimeter is 2 times the sum of the width and length.

$P = 2(3 \text{ m} + 5 \text{ m})$

Step 2   Find the sum of the width and length.

$P = 2(8 \text{ m})$

Step 3   Multiply by 2.

$P = 16 \text{ m}$

The perimeter is 16 m.

Example 2

Find the perimeter of a shape with sides measuring 2 cm, 5 cm, 6 cm, 3 cm.

Step 1   You know that the perimeter is the sum of all the sides.

$P = 2 + 5 + 6 + 3$

Step 2   Find the sum of the sides.

$P = 2 + 5 + 6 + 3$

$P = 16$

The perimeter is 16 cm.

Practice Problem Find the perimeter of a rectangle with a length of 18 m and a width of 7 m.

Practice Problem Find the perimeter of a triangle measuring 1.6 cm by 2.4 cm by 2.4 cm.

Area of a Rectangle The **area** (A) is the number of square units needed to cover a surface. To find the area of a rectangle, multiply the length times the width, or $l \times w$. When finding area, the units also are multiplied. Area is given in square units.

Example

Find the area of a rectangle with a length of 1 cm and a width of 10 cm.

Step 1   You know that the area is the length multiplied by the width.

$A = (1 \text{ cm} \times 10 \text{ cm})$

Step 2   Multiply the length by the width. Also multiply the units.

$A = 10 \text{ cm}^2$

The area is 10 cm^2.

Practice Problem Find the area of a square whose sides measure 4 m.

Area of a Triangle To find the area of a triangle, use the formula:

$A = \frac{1}{2}(\text{base} \times \text{height})$

The base of a triangle can be any of its sides. The height is the perpendicular distance from a base to the opposite endpoint, or vertex.

Example

Find the area of a triangle with a base of 18 m and a height of 7 m.

Step 1   You know that the area is $\frac{1}{2}$ the base times the height.

$A = \frac{1}{2}(18 \text{ m} \times 7 \text{ m})$

Step 2   Multiply $\frac{1}{2}$ by the product of 18×7. Multiply the units.

$A = \frac{1}{2}(126 \text{ m}^2)$

$A = 63 \text{ m}^2$

The area is 63 m^2.

Practice Problem Find the area of a triangle with a base of 27 cm and a height of 17 cm.

SCIENCE SKILL HANDBOOK

MATH SKILL HANDBOOK

FOLDABLES HANDBOOK

REFERENCE HANDBOOK

GLOSSARY/ GLOSARIO

INDEX

SCIENCE SKILL HANDBOOK

MATH SKILL HANDBOOK

FOLDABLES HANDBOOK

REFERENCE HANDBOOK

GLOSSARY/ GLOSARIO

INDEX

Circumference of a Circle The **diameter** (*d*) of a circle is the distance across the circle through its center, and the **radius** (r) is the distance from the center to any point on the circle. The radius is half of the diameter. The distance around the circle is called the **circumference** (C). The formula for finding the circumference is:

$$C = 2\pi r \text{ or } C = \pi d$$

The circumference divided by the diameter is always equal to 3.1415926… This nonterminating and nonrepeating number is represented by the Greek letter π (pi). An approximation often used for π is 3.14.

Example 1

Find the circumference of a circle with a radius of 3 m.

Step 1 You know the formula for the circumference is 2 times the radius times π.

$$C = 2\pi(3)$$

Step 2 Multiply 2 times the radius.

$$C = 6\pi$$

Step 3 Multiply by π.

$$C \approx 19 \text{ m}$$

The circumference is about 19 m.

Example 2

Find the circumference of a circle with a diameter of 24.0 cm.

Step 1 You know the formula for the circumference is the diameter times π.

$$C = \pi(24.0)$$

Step 2 Multiply the diameter by π.

$$C \approx 75.4 \text{ cm}$$

The circumference is about 75.4 cm.

Practice Problem Find the circumference of a circle with a radius of 19 cm.

Area of a Circle The formula for the area of a circle is: $A = \pi r^2$

Example 1

Find the area of a circle with a radius of 4.0 cm.

Step 1 $A = \pi(4.0)^2$

Step 2 Find the square of the radius.

$$A = 16\pi$$

Step 3 Multiply the square of the radius by π.

$$A \approx 50 \text{ cm}^2$$

The area of the circle is about 50 cm^2.

Example 2

Find the area of a circle with a radius of 225 m.

Step 1 $A = \pi(225)^2$

Step 2 Find the square of the radius.

$$A = 50625\pi$$

Step 3 Multiply the square of the radius by π.

$$A \approx 159043.1$$

The area of the circle is about 159043.1 m^2.

Example 3

Find the area of a circle whose diameter is 20.0 mm.

Step 1 Remember that the radius is half of the diameter.

$$A = \pi\left(\frac{20.0}{2}\right)^2$$

Step 2 Find the radius.

$$A = \pi(10.0)^2$$

Step 3 Find the square of the radius.

$$A = 100\pi$$

Step 4 Multiply the square of the radius by π.

$$A \approx 314 \text{ mm}^2$$

The area of the circle is about 314 mm^2.

Practice Problem Find the area of a circle with a radius of 16 m.

Volume The measure of space occupied by a solid is the **volume** (V). To find the volume of a rectangular solid multiply the length times width times height, or $V = l \times w \times h$. It is measured in cubic units, such as cubic centimeters (cm^3).

Example

Find the volume of a rectangular solid with a length of 2.0 m, a width of 4.0 m, and a height of 3.0 m.

Step 1 You know the formula for volume is the length times the width times the height.

$$V = 2.0 \text{ m} \times 4.0 \text{ m} \times 3.0 \text{ m}$$

Step 2 Multiply the length times the width times the height.

$$V = 24 \text{ m}^3$$

The volume is 24 m^3.

Practice Problem Find the volume of a rectangular solid that is 8 m long, 4 m wide, and 4 m high.

To find the volume of other solids, multiply the area of the base times the height.

Example 1

Find the volume of a solid that has a triangular base with a length of 8.0 m and a height of 7.0 m. The height of the entire solid is 15.0 m.

Step 1 You know that the base is a triangle, and the area of a triangle is $\frac{1}{2}$ the base times the height, and the volume is the area of the base times the height.

$$V = \left[\frac{1}{2} (b \times h)\right] \times 15$$

Step 2 Find the area of the base.

$$V = \left[\frac{1}{2} (8 \times 7)\right] \times 15$$

$$V = \left(\frac{1}{2} \times 56\right) \times 15$$

Step 3 Multiply the area of the base by the height of the solid.

$$V = 28 \times 15$$

$$V = 420 \text{ m}^3$$

The volume is 420 m^3.

Example 2

Find the volume of a cylinder that has a base with a radius of 12.0 cm, and a height of 21.0 cm.

Step 1 You know that the base is a circle, and the area of a circle is the square of the radius times π, and the volume is the area of the base times the height.

$$V = (\pi r^2) \times 21$$

$$V = (\pi 12^2) \times 21$$

Step 2 Find the area of the base.

$$V = 144\pi \times 21$$

$$V = 452 \times 21$$

Step 3 Multiply the area of the base by the height of the solid.

$$V \approx 9{,}500 \text{ cm}^3$$

The volume is about 9,500 cm^3.

Example 3

Find the volume of a cylinder that has a diameter of 15 mm and a height of 4.8 mm.

Step 1 You know that the base is a circle with an area equal to the square of the radius times π. The radius is one-half the diameter. The volume is the area of the base times the height.

$$V = (\pi r^2) \times 4.8$$

$$V = \left[\pi\left(\frac{1}{2} \times 15\right)^2\right] \times 4.8$$

$$V = (\pi 7.5^2) \times 4.8$$

Step 2 Find the area of the base.

$$V = 56.25\pi \times 4.8$$

$$V \approx 176.71 \times 4.8$$

Step 3 Multiply the area of the base by the height of the solid.

$$V \approx 848.2$$

The volume is about 848.2 mm^3.

Practice Problem Find the volume of a cylinder with a diameter of 7 cm in the base and a height of 16 cm.

SCIENCE SKILL HANDBOOK

MATH SKILL HANDBOOK

FOLDABLES HANDBOOK

REFERENCE HANDBOOK

GLOSSARY/ GLOSARIO

INDEX

Science Applications

SCIENCE SKILL HANDBOOK

MATH SKILL HANDBOOK

FOLDABLES HANDBOOK

REFERENCE HANDBOOK

GLOSSARY/ GLOSARIO

INDEX

Measure in SI

The metric system of measurement was developed in 1795. A modern form of the metric system, called the International System (SI), was adopted in 1960 and provides the standard measurements that all scientists around the world can understand.

The SI system is convenient because unit sizes vary by powers of 10. Prefixes are used to name units. Look at **Table 2** for some common SI prefixes and their meanings.

Table 2 Common SI Prefixes			
Prefix	Symbol	Meaning	
kilo–	k	1,000	thousandth
hecto–	h	100	hundred
deka–	da	10	ten
deci–	d	0.1	tenth
centi–	c	0.01	hundreth
milli–	m	0.001	thousandth

Example

How many grams equal one kilogram?

Step 1 Find the prefix *kilo–* in **Table 2.**

Step 2 Using **Table 2,** determine the meaning of *kilo–*. According to the table, it means 1,000. When the prefix *kilo–* is added to a unit, it means that there are 1,000 of the units in a "kilounit."

Step 3 Apply the prefix to the units in the question. The units in the question are grams. There are 1,000 grams in a kilogram.

Practice Problem Is a milligram larger or smaller than a gram? How many of the smaller units equal one larger unit? What fraction of the larger unit does one smaller unit represent?

Dimensional Analysis

Convert SI Units In science, quantities such as length, mass, and time sometimes are measured using different units. A process called dimensional analysis can be used to change one unit of measure to another. This process involves multiplying your starting quantity and units by one or more conversion factors. A conversion factor is a ratio equal to one and can be made from any two equal quantities with different units. If 1,000 mL equal 1 L then two ratios can be made.

$$\frac{1{,}000 \text{ mL}}{1 \text{ L}} = \frac{1 \text{ L}}{1{,}000 \text{ mL}} = 1$$

One can convert between units in the SI system by using the equivalents in **Table 2** to make conversion factors.

Example

How many cm are in 4 m?

Step 1 Write conversion factors for the units given. From **Table 2,** you know that 100 cm = 1 m. The conversion factors are

$$\frac{100 \text{ cm}}{1 \text{ m}} \text{ and } \frac{1 \text{ m}}{100 \text{ cm}}$$

Step 2 Decide which conversion factor to use. Select the factor that has the units you are converting from (m) in the denominator and the units you are converting to (cm) in the numerator.

$$\frac{100 \text{ cm}}{1 \text{ m}}$$

Step 3 Multiply the starting quantity and units by the conversion factor. Cancel the starting units with the units in the denominator. There are 400 cm in 4 m.

$$4 \text{ m} = \frac{100 \text{ cm}}{1 \text{ m}} = 400 \text{ cm}$$

Practice Problem How many milligrams are in one kilogram? (Hint: You will need to use two conversion factors from **Table 2.**)

Table 3 Unit System Equivalents

Type of Measurement	Equivalent
Length	1 in = 2.54 cm 1 yd = 0.91 m 1 mi = 1.61 km
Mass and weight*	1 oz = 28.35 g 1 lb = 0.45 kg 1 ton (short) = 0.91 tonnes (metric tons) 1 lb = 4.45 N
Volume	$1\ in^3 = 16.39\ cm^3$ 1 qt = 0.95 L 1 gal = 3.78 L
Area	$1\ in^2 = 6.45\ cm^2$ $1\ yd^2 = 0.83\ m^2$ $1\ mi^2 = 2.59\ km^2$ 1 acre = 0.40 hectares
Temperature	$°C = \frac{(°F - 32)}{1.8}$ K = °C + 273

*Weight is measured in standard Earth gravity.

Convert Between Unit Systems **Table 3** gives a list of equivalents that can be used to convert between English and SI units.

Example

If a meterstick has a length of 100 cm, how long is the meterstick in inches?

Step 1 Write the conversion factors for the units given. From **Table 3,** 1 in = 2.54 cm.

$$\frac{1\ in}{2.54\ cm}\ and\ \frac{2.54\ cm}{1\ in}$$

Step 2 Determine which conversion factor to use. You are converting from cm to in. Use the conversion factor with cm on the bottom.

$$\frac{1\ in}{2.54\ cm}$$

Step 3 Multiply the starting quantity and units by the conversion factor. Cancel the starting units with the units in the denominator. Round your answer to the nearest tenth.

$$100\ cm \times \frac{1\ in}{2.54\ cm} = 39.37\ in$$

The meterstick is about 39.4 in long.

Practice Problem 1 A book has a mass of 5 lb. What is the mass of the book in kg?

Practice Problem 2 Use the equivalent for in and cm (1 in = 2.54 cm) to show how $1\ in^3 \approx 16.39\ cm^3$.

SCIENCE SKILL HANDBOOK

MATH SKILL HANDBOOK

FOLDABLES HANDBOOK

REFERENCE HANDBOOK

GLOSSARY/ GLOSARIO

INDEX

SCIENCE SKILL HANDBOOK

MATH SKILL HANDBOOK

FOLDABLES HANDBOOK

REFERENCE HANDBOOK

GLOSSARY/ GLOSARIO

INDEX

Precision and Significant Digits

When you make a measurement, the value you record depends on the precision of the measuring instrument. This precision is represented by the number of significant digits recorded in the measurement. When counting the number of significant digits, all digits are counted except zeros at the end of a number with no decimal point such as 2,050, and zeros at the beginning of a decimal such as 0.03020. When adding or subtracting numbers with different precision, round the answer to the smallest number of decimal places of any number in the sum or difference. When multiplying or dividing, the answer is rounded to the smallest number of significant digits of any number being multiplied or divided.

Example

The lengths 5.28 and 5.2 are measured in meters. Find the sum of these lengths and record your answer using the correct number of significant digits.

Step 1 Find the sum.

5.28 m	2 digits after the decimal
+ 5.2 m	1 digit after the decimal
10.48 m	

Step 2 Round to one digit after the decimal because the least number of digits after the decimal of the numbers being added is 1.

The sum is 10.5 m.

Practice Problem 1 How many significant digits are in the measurement 7,071,301 m? How many significant digits are in the measurement 0.003010 g?

Practice Problem 2 Multiply 5.28 and 5.2 using the rule for multiplying and dividing. Record the answer using the correct number of significant digits.

Scientific Notation

Many times numbers used in science are very small or very large. Because these numbers are difficult to work with scientists use scientific notation. To write numbers in scientific notation, move the decimal point until only one non-zero digit remains on the left. Then count the number of places you moved the decimal point and use that number as a power of ten. For example, the average distance from the Sun to Mars is 227,800,000,000 m. In scientific notation, this distance is 2.278×10^{11} m. Because you moved the decimal point to the left, the number is a positive power of ten.

The mass of an electron is about 0.000 000 000 000 000 000 000 000 000 000 911 kg. Expressed in scientific notation, this mass is 9.11×10^{-31} kg. Because the decimal point was moved to the right, the number is a negative power of ten.

Example

Earth is 149,600,000 km from the Sun. Express this in scientific notation.

Step 1 Move the decimal point until one non-zero digit remains on the left.

1.496 000 00

Step 2 Count the number of decimal places you have moved. In this case, eight.

Step 2 Show that number as a power of ten, 10^8.

Earth is 1.496×10^8 km from the Sun.

Practice Problem 1 How many significant digits are in 149,600,000 km? How many significant digits are in 1.496×10^8 km?

Practice Problem 2 Parts used in a high performance car must be measured to 7×10^{-6} m. Express this number as a decimal.

Practice Problem 3 A CD is spinning at 539 revolutions per minute. Express this number in scientific notation.

Make and Use Graphs

Data in tables can be displayed in a graph—a visual representation of data. Common graph types include line graphs, bar graphs, and circle graphs.

Line Graph A line graph shows a relationship between two variables that change continuously. The independent variable is changed and is plotted on the x-axis. The dependent variable is observed, and is plotted on the y-axis.

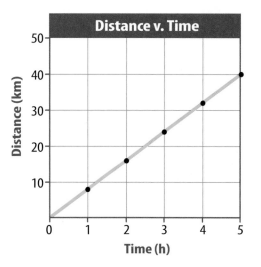

Figure 8 This line graph shows the relationship between distance and time during a bicycle ride.

Practice Problem A puppy's shoulder height is measured during the first year of her life. The following measurements were collected: (3 mo, 52 cm), (6 mo, 72 cm), (9 mo, 83 cm), (12 mo, 86 cm). Graph this data.

Find a Slope The slope of a straight line is the ratio of the vertical change, rise, to the horizontal change, run.

$$\text{Slope} = \frac{\text{vertical change (rise)}}{\text{horizontal change (run)}} = \frac{\text{change in } y}{\text{change in } x}$$

Example

Draw a line graph of the data below from a cyclist in a long-distance race.

Table 4 Bicycle Race Data	
Time (h)	**Distance (km)**
0	0
1	8
2	16
3	24
4	32
5	40

Step 1 Determine the x-axis and y-axis variables. Time varies independently of distance and is plotted on the x-axis. Distance is dependent on time and is plotted on the y-axis.

Step 2 Determine the scale of each axis. The x-axis data ranges from 0 to 5. The y-axis data ranges from 0 to 50.

Step 3 Using graph paper, draw and label the axes. Include units in the labels.

Step 4 Draw a point at the intersection of the time value on the x-axis and corresponding distance value on the y-axis. Connect the points and label the graph with a title, as shown in **Figure 8.**

Example

Find the slope of the graph in **Figure 8**.

Step 1 You know that the slope is the change in y divided by the change in x.

$$\text{Slope} = \frac{\text{change in } y}{\text{change in } x}$$

Step 2 Determine the data points you will be using. For a straight line, choose the two sets of points that are the farthest apart.

$$\text{Slope} = \frac{(40 - 0) \text{ km}}{(5 - 0) \text{ h}}$$

Step 3 Find the change in y and x.

$$\text{Slope} = \frac{40 \text{ km}}{5 \text{ h}}$$

Step 4 Divide the change in y by the change in x.

$$\text{Slope} = \frac{8 \text{ km}}{\text{h}}$$

The slope of the graph is 8 km/h.

SCIENCE SKILL HANDBOOK

MATH SKILL HANDBOOK

FOLDABLES HANDBOOK

REFERENCE HANDBOOK

GLOSSARY/ GLOSARIO

INDEX

SCIENCE SKILL HANDBOOK

MATH SKILL HANDBOOK

FOLDABLES HANDBOOK

REFERENCE HANDBOOK

GLOSSARY/ GLOSARIO

INDEX

Bar Graph To compare data that does not change continuously you might choose a bar graph. A bar graph uses bars to show the relationships between variables. The *x*-axis variable is divided into parts. The parts can be numbers such as years, or a category such as a type of animal. The *y*-axis is a number and increases continuously along the axis.

Example

A recycling center collects 4.0 kg of aluminum on Monday, 1.0 kg on Wednesday, and 2.0 kg on Friday. Create a bar graph of this data.

Step 1 Select the *x*-axis and *y*-axis variables. The measured numbers (the masses of aluminum) should be placed on the *y*-axis. The variable divided into parts (collection days) is placed on the *x*-axis.

Step 2 Create a graph grid like you would for a line graph. Include labels and units.

Step 3 For each measured number, draw a vertical bar above the *x*-axis value up to the *y*-axis value. For the first data point, draw a vertical bar above Monday up to 4.0 kg.

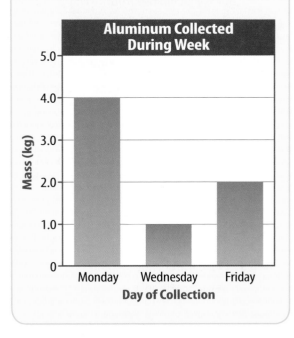

Practice Problem Draw a bar graph of the gases in air: 78% nitrogen, 21% oxygen, 1% other gases.

Circle Graph To display data as parts of a whole, you might use a circle graph. A circle graph is a circle divided into sections that represent the relative size of each piece of data. The entire circle represents 100%, half represents 50%, and so on.

Example

Air is made up of 78% nitrogen, 21% oxygen, and 1% other gases. Display the composition of air in a circle graph.

Step 1 Multiply each percent by 360° and divide by 100 to find the angle of each section in the circle.

$$78\% \times \frac{360°}{100} = 280.8°$$

$$21\% \times \frac{360°}{100} = 75.6°$$

$$1\% \times \frac{360°}{100} = 3.6°$$

Step 2 Use a compass to draw a circle and to mark the center of the circle. Draw a straight line from the center to the edge of the circle.

Step 3 Use a protractor and the angles you calculated to divide the circle into parts. Place the center of the protractor over the center of the circle and line the base of the protractor over the straight line.

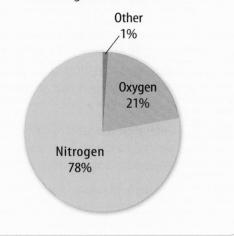

Practice Problem Draw a circle graph to represent the amount of aluminum collected during the week shown in the bar graph to the left.

Student Study Guides & Instructions
By Dinah Zike

1. You will find suggestions for Study Guides, also known as Foldables or books, in each chapter lesson and as a final project. Look at the end of the chapter to determine the project format and glue the Foldables in place as you progress through the chapter lessons.

2. Creating the Foldables or books is simple and easy to do by using copy paper, art paper, and internet printouts. Photocopies of maps, diagrams, or your own illustrations may also be used for some of the Foldables. Notebook paper is the most common source of material for study guides and 83% of all Foldables are created from it. When folded to make books, notebook paper Foldables easily fit into 11" × 17" or 12" × 18" chapter projects with space left over. Foldables made using photocopy paper are slightly larger and they fit into Projects, but snugly. Use the least amount of glue, tape, and staples needed to assemble the Foldables.

3. Seven of the Foldables can be made using either small or large paper. When 11" × 17" or 12" × 18" paper is used, these become projects for housing smaller Foldables. Project format boxes are located within the instructions to remind you of this option.

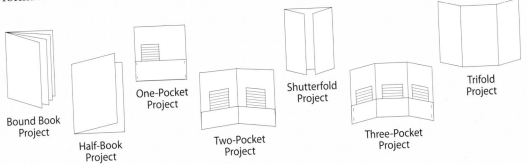

Bound Book Project

Half-Book Project

One-Pocket Project

Two-Pocket Project

Shutterfold Project

Three-Pocket Project

Trifold Project

4. Use one-gallon self-locking plastic bags to store your projects. Place strips of two-inch clear tape along the left, long side of the bag and punch holes through the taped edge. Cut the bottom corners off the bag so it will not hold air. Store this Project Portfolio inside a three-hole binder. To store a large collection of project bags, use a giant laundry-soap box. Holes can be punched in some of the Foldable Projects so they can be stored in a three-hole binder without using a plastic bag. Punch holes in the pocket books before gluing or stapling the pocket.

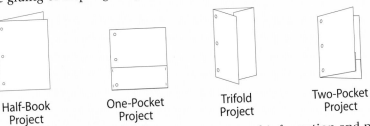

Half-Book Project

One-Pocket Project

Trifold Project

Two-Pocket Project

5. Maximize the use of the projects by collecting additional information and placing it on the back of the project and other unused spaces of the large Foldables.

SCIENCE SKILL HANDBOOK

MATH SKILL HANDBOOK

FOLDABLES HANDBOOK

REFERENCE HANDBOOK

GLOSSARY/ GLOSARIO

INDEX

Half-Book Foldable® By Dinah Zike

Step 1 Fold a sheet of notebook or copy paper in half.

Label the exterior tab and use the inside space to write information.

Variations

Paper can be folded horizontally, like a *hamburger* or vertically, like a *hot dog*.

C Half-books can be folded so that one side is ½ inch longer than the other side. A title or question can be written on the extended tab.

PROJECT FORMAT
Use 11" × 17" or 12" × 18" paper on the horizontal axis to make a large project book.

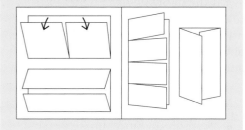

Worksheet Foldable or Folded Book® By Dinah Zike

Step 1 Make a half-book (see above) using work sheets, internet print-outs, diagrams, or maps.

Step 2 Fold it in half again.

Variations

A This folded sheet as a small book with two pages can be used for comparing and contrasting, cause and effect, or other skills.

B When the sheet of paper is open, the four sections can be used separately or used collectively to show sequences or steps.

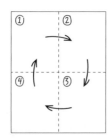

Science Skill Handbook

Math Skill Handbook

Foldables Handbook

Reference Handbook

Glossary/Glosario

Index

Two-Tab and Concept-Map Foldable® By Dinah Zike

Step 1 Fold a sheet of notebook or copy paper in half vertically or horizontally.

Step 2 Fold it in half again, as shown.

Step 3 Unfold once and cut along the fold line or valley of the top flap to make two flaps.

Variations

A Concept maps can be made by leaving a ½ inch tab at the top when folding the paper in half. Use arrows and labels to relate topics to the primary concept.

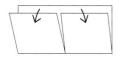

B Use two sheets of paper to make multiple page tab books. Glue or staple books together at the top fold.

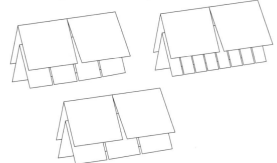

• •

Three-Quarter Foldable® By Dinah Zike

Step 1 Make a two-tab book (see above) and cut the left tab off at the top of the fold line.

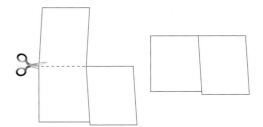

Variations

A Use this book to draw a diagram or a map on the exposed left tab. Write questions about the illustration on the top right tab and provide complete answers on the space under the tab.

B Compose a self-test using multiple choice answers for your questions. Include the correct answer with three wrong responses. The correct answers can be written on the back of the book or upside down on the bottom of the inside page.

SCIENCE SKILL HANDBOOK

MATH SKILL HANDBOOK

FOLDABLES HANDBOOK

REFERENCE HANDBOOK

GLOSSARY/ GLOSARIO

INDEX

SCIENCE SKILL HANDBOOK

MATH SKILL HANDBOOK

FOLDABLES HANDBOOK

REFERENCE HANDBOOK

GLOSSARY/ GLOSARIO

INDEX

Three-Tab Foldable® By Dinah Zike

Step 1 Fold a sheet of paper in half horizontally.

Step 2 Fold into thirds.

Step 3 Unfold and cut along the folds of the top flap to make three sections.

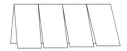

Variations

A Before cutting the three tabs draw a Venn diagram across the front of the book.

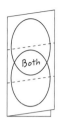

B Make a space to use for titles or concept maps by leaving a ½ inch tab at the top when folding the paper in half.

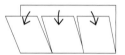

Four-Tab Foldable® By Dinah Zike

Step 1 Fold a sheet of paper in half horizontally.

Step 2 Fold in half and then fold each half as shown below.

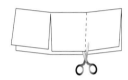

Step 3 Unfold and cut along the fold lines of the top flap to make four tabs.

Variations

A Make a space to use for titles or concept maps by leaving a ½ inch tab at the top when folding the paper in half.

B Use the book on the vertical axis, with or without an extended tab.

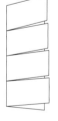

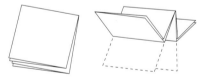

Folding Fifths for a Foldable® By Dinah Zike

SCIENCE SKILL HANDBOOK

MATH SKILL HANDBOOK

FOLDABLES HANDBOOK

REFERENCE HANDBOOK

GLOSSARY/ GLOSARIO

INDEX

Step 1 Fold a sheet of paper in half horizontally.

Step 2 Fold again so one-third of the paper is exposed and two-thirds are covered.

Step 3 Fold the two-thirds section in half.

Step 4 Fold the one-third section, a single thickness, backward to make a fold line.

Variations

A Unfold and cut along the fold lines to make five tabs.

B Make a five-tab book with a ½ inch tab at the top (see two-tab instructions).

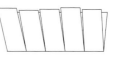

C Use 11″ × 17″ or 12″ × 18″ paper and fold into fifths for a five-column and/or row table or chart.

- -

Folded Table or Chart, and Trifold Foldable® By Dinah Zike

Step 1 Fold a sheet of paper in the required number of vertical columns for the table or chart.

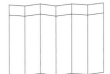

Step 2 Fold the horizontal rows needed for the table or chart.

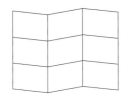

PROJECT FORMAT
Use 11″ × 17″ or 12″ × 18″ paper and fold it to make a large trifold project book or larger tables and charts.

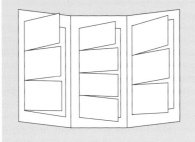

Variations

A Make a trifold by folding the paper into thirds vertically or horizontally.

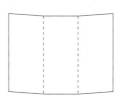

B Make a trifold book. Unfold it and draw a Venn diagram on the inside.

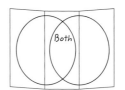

Two or Three-Pockets Foldable® By Dinah Zike

Step 1 Fold up the long side of a horizontal sheet of paper about 5 cm.

Step 2 Fold the paper in half.

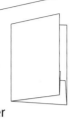

Step 3 Open the paper and glue or staple the outer edges to make two compartments.

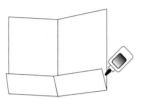

Variations

A Make a multi-page booklet by gluing several pocket books together.

B Make a three-pocket book by using a trifold (see previous instructions).

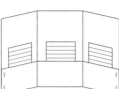

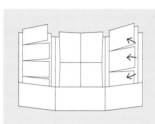

PROJECT FORMAT
Use 11" × 17" or 12" × 18" paper and fold it horizontally to make a large multi-pocket project.

Matchbook Foldable® By Dinah Zike

Step 1 Fold a sheet of paper almost in half and make the back edge about 1–2 cm longer than the front edge.

Step 2 Find the midpoint of the shorter flap.

Step 3 Open the paper and cut the short side along the midpoint making two tabs.

Step 4 Close the book and fold the tab over the short side.

Variations

A Make a single-tab matchbook by skipping Steps 2 and 3.

B Make two smaller matchbooks by cutting the single-tab matchbook in half.

SCIENCE SKILL HANDBOOK

MATH SKILL HANDBOOK

FOLDABLES HANDBOOK

REFERENCE HANDBOOK

GLOSSARY/GLOSARIO

INDEX

Shutterfold Foldable® By Dinah Zike

Step 1 Begin as if you were folding a vertical sheet of paper in half, but instead of creasing the paper, pinch it to show the midpoint.

Step 2 Fold the top and bottom to the middle and crease the folds.

Variations

A Use the shutterfold on the horizontal axis.

B Create a center tab by leaving .5–2 cm between the flaps in Step 2.

PROJECT FORMAT
Use 11″ × 17″ or 12″ × 18″ paper and fold it to make a large shutterfold project.

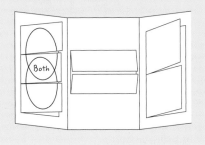

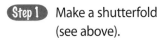

Four-Door Foldable® By Dinah Zike

Step 1 Make a shutterfold (see above).

Step 2 Fold the sheet of paper in half.

Step 3 Open the last fold and cut along the inside fold lines to make four tabs.

Variations

A Use the four-door book on the opposite axis.

B Create a center tab by leaving .5–2 cm between the flaps in Step 1.

SCIENCE SKILL HANDBOOK

MATH SKILL HANDBOOK

FOLDABLES HANDBOOK

REFERENCE HANDBOOK

GLOSSARY/ GLOSARIO

INDEX

Bound Book Foldable® By Dinah Zike

Step 1 Fold three sheets of paper in half. Place the papers in a stack, leaving about .5 cm between each top fold. Mark all three sheets about 3 cm from the outer edges.

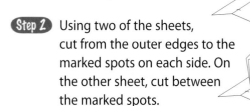

Step 2 Using two of the sheets, cut from the outer edges to the marked spots on each side. On the other sheet, cut between the marked spots.

Step 3 Take the two sheets from Step 1 and slide them through the cut in the third sheet to make a 12-page book.

Step 4 Fold the bound pages in half to form a book.

Variation

A Use two sheets of paper to make an eight-page book, or increase the number of pages by using more than three sheets.

PROJECT FORMAT
Use two or more sheets of 11″ × 17″ or 12″ × 18″ paper and fold it to make a large bound book project.

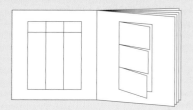

Accordian Foldable® By Dinah Zike

Step 1 Fold the selected paper in half vertically, like a *hamburger*.

Step 2 Cut each sheet of folded paper in half along the fold lines.

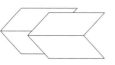

Step 3 Fold each half-sheet almost in half, leaving a 2 cm tab at the top.

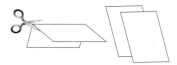

Step 4 Fold the top tab over the short side, then fold it in the opposite direction.

Variations

A Glue the straight edge of one paper inside the tab of another sheet. Leave a tab at the end of the book to add more pages.

B Tape the straight edge of one paper to the tab of another sheet, or just tape the straight edges of nonfolded paper end to end to make an accordian.

C Use whole sheets of paper to make a large accordian.

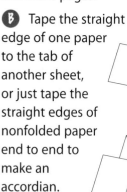

SCIENCE SKILL HANDBOOK

MATH SKILL HANDBOOK

FOLDABLES HANDBOOK

REFERENCE HANDBOOK

GLOSSARY/ GLOSARIO

INDEX

Layered Foldable® By Dinah Zike

Step 1 Stack two sheets of paper about 1–2 cm apart. Keep the right and left edges even.

Step 2 Fold up the bottom edges to form four tabs. Crease the fold to hold the tabs in place.

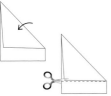

Step 3 Staple along the folded edge, or open and glue the papers together at the fold line.

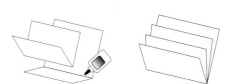

Variations

A Rotate the book so the fold is at the top or to the side.

B Extend the book by using more than two sheets of paper.

Envelope Foldable® By Dinah Zike

Step 1 Fold a sheet of paper into a *taco*. Cut off the tab at the top.

Step 2 Open the *taco* and fold it the opposite way making another *taco* and an X-fold pattern on the sheet of paper.

Step 3 Cut a map, illustration, or diagram to fit the inside of the envelope.

Step 4 Use the outside tabs for labels and inside tabs for writing information.

Variations

A Use 11″ × 17″ or 12″ × 18″ paper to make a large envelope.

B Cut off the points of the four tabs to make a window in the middle of the book.

SCIENCE SKILL HANDBOOK

MATH SKILL HANDBOOK

FOLDABLES HANDBOOK

REFERENCE HANDBOOK

GLOSSARY/ GLOSARIO

INDEX

Sentence Strip Foldable® By Dinah Zike

Step 1 Fold two sheets of paper in half vertically, like a *hamburger*.

Step 2 Unfold and cut along fold lines making four half sheets.

Step 3 Fold each half sheet in half horizontally, like a *hot dog*.

Step 4 Stack folded horizontal sheets evenly and staple together on the left side.

Step 5 Open the top flap of the first sentence strip and make a cut about 2 cm from the stapled edge to the fold line. This forms a flap that can be raisied and lowered. Repeat this step for each sentence strip.

Variations

A Expand this book by using more than two sheets of paper.

B Use whole sheets of paper to make large books.

Pyramid Foldable® By Dinah Zike

Step 1 Fold a sheet of paper into a *taco*. Crease the fold line, but do not cut it off.

Step 2 Open the folded sheet and refold it like a *taco* in the opposite direction to create an X-fold pattern.

Step 3 Cut one fold line as shown, stopping at the center of the X-fold to make a flap.

Step 4 Outline the fold lines of the X-fold. Label the three front sections and use the inside spaces for notes. Use the tab for the title.

Step 5 Glue the tab into a project book or notebook. Use the space under the pyramid for other information.

Step 6 To display the pyramid, fold the flap under and secure with a paper clip, if needed.

Single-Pocket or One-Pocket Foldable® By Dinah Zike

Step 1 Using a large piece of paper on a vertical axis, fold the bottom edge of the paper upwards, about 5 cm.

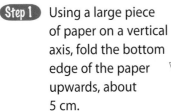

Step 2 Glue or staple the outer edges to make a large pocket.

PROJECT FORMAT
Use 11″ × 17″ or 12″ × 18″ paper and fold it vertically or horizontally to make a large pocket project.

Variations

A Make the one-pocket project using the paper on the horizontal axis.

B To store materials securely inside, fold the top of the paper almost to the center, leaving about 2–4 cm between the paper edges. Slip the Foldables through the opening and under the top and bottom pockets.

Multi-Tab Foldable® By Dinah Zike

Step 1 Fold a sheet of notebook paper in half like a *hot dog*.

Step 2 Open the paper and on one side cut every third line. This makes ten tabs on wide ruled notebook paper and twelve tabs on college ruled.

Step 3 Label the tabs on the front side and use the inside space for definitions or other information.

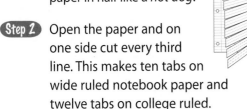

Variation

A Make a tab for a title by folding the paper so the holes remain uncovered. This allows the notebook Foldable to be stored in a three-hole binder.

SCIENCE SKILL HANDBOOK

MATH SKILL HANDBOOK

FOLDABLES HANDBOOK

REFERENCE HANDBOOK

GLOSSARY/ GLOSARIO

INDEX

PERIODIC TABLE OF THE ELEMENTS

Element — Hydrogen
Atomic number — 1
Symbol — **H**
Atomic mass — 1.01

State of matter

🎈 Gas
💧 Liquid
☐ Solid
⊙ Synthetic

A column in the periodic table is called a **group.**

A row in the periodic table is called a **period.**

1

	1	2		3	4	5	6	7	8	9
1	Hydrogen 1 **H** 1.01									
2	Lithium 3 **Li** 6.94	Beryllium 4 **Be** 9.01								
3	Sodium 11 **Na** 22.99	Magnesium 12 **Mg** 24.31								
4	Potassium 19 **K** 39.10	Calcium 20 **Ca** 40.08	Scandium 21 **Sc** 44.96	Titanium 22 **Ti** 47.87	Vanadium 23 **V** 50.94	Chromium 24 **Cr** 52.00	Manganese 25 **Mn** 54.94	Iron 26 **Fe** 55.85	Cobalt 27 **Co** 58.93
5	Rubidium 37 **Rb** 85.47	Strontium 38 **Sr** 87.62	Yttrium 39 **Y** 88.91	Zirconium 40 **Zr** 91.22	Niobium 41 **Nb** 92.91	Molybdenum 42 **Mo** 95.96	Technetium 43 **Tc** (98)	Ruthenium 44 **Ru** 101.07	Rhodium 45 **Rh** 102.91
6	Cesium 55 **Cs** 132.91	Barium 56 **Ba** 137.33	Lanthanum 57 **La** 138.91	Hafnium 72 **Hf** 178.49	Tantalum 73 **Ta** 180.95	Tungsten 74 **W** 183.84	Rhenium 75 **Re** 186.21	Osmium 76 **Os** 190.23	Iridium 77 **Ir** 192.22
7	Francium 87 **Fr** (223)	Radium 88 **Ra** (226)	Actinium 89 **Ac** (227)	Rutherfordium 104 **Rf** (267)	Dubnium 105 **Db** (268)	Seaborgium 106 **Sg** (271)	Bohrium 107 **Bh** (272)	Hassium 108 **Hs** (270)	Meitnerium 109 **Mt** (276)

The number in parentheses is the mass number of the longest lived isotope for that element.

Lanthanide series

Cerium 58 **Ce** 140.12	Praseodymium 59 **Pr** 140.91	Neodymium 60 **Nd** 144.24	Promethium 61 **Pm** (145)	Samarium 62 **Sm** 150.36	Europium 63 **Eu** 151.96

Actinide series

Thorium 90 **Th** 232.04	Protactinium 91 **Pa** 231.04	Uranium 92 **U** 238.03	Neptunium 93 **Np** (237)	Plutonium 94 **Pu** (244)	Americium 95 **Am** (243)

SCIENCE SKILL HANDBOOK
MATH SKILL HANDBOOK
FOLDABLES HANDBOOK
REFERENCE HANDBOOK
GLOSSARY/GLOSARIO
INDEX

Metal
Metalloid
Nonmetal
Recently discovered

			13	14	15	16	17	18
								Helium 2 He 4.00
			Boron 5 B 10.81	Carbon 6 C 12.01	Nitrogen 7 N 14.01	Oxygen 8 O 16.00	Fluorine 9 F 19.00	Neon 10 Ne 20.18
10	11	12	Aluminum 13 Al 26.98	Silicon 14 Si 28.09	Phosphorus 15 P 30.97	Sulfur 16 S 32.07	Chlorine 17 Cl 35.45	Argon 18 Ar 39.95
Nickel 28 Ni 58.69	Copper 29 Cu 63.55	Zinc 30 Zn 65.38	Gallium 31 Ga 69.72	Germanium 32 Ge 72.64	Arsenic 33 As 74.92	Selenium 34 Se 78.96	Bromine 35 Br 79.90	Krypton 36 Kr 83.80
Palladium 46 Pd 106.42	Silver 47 Ag 107.87	Cadmium 48 Cd 112.41	Indium 49 In 114.82	Tin 50 Sn 118.71	Antimony 51 Sb 121.76	Tellurium 52 Te 127.60	Iodine 53 I 126.90	Xenon 54 Xe 131.29
Platinum 78 Pt 195.08	Gold 79 Au 196.97	Mercury 80 Hg 200.59	Thallium 81 Tl 204.38	Lead 82 Pb 207.20	Bismuth 83 Bi 208.98	Polonium 84 Po (209)	Astatine 85 At (210)	Radon 86 Rn (222)
Darmstadtium 110 Ds (281)	Roentgenium 111 Rg (280)	Copernicium 112 Cn (285)	Ununtrium * 113 Uut (284)	Ununquadium * 114 Uuq (289)	Ununpentium * 115 Uup (288)	Ununhexium * 116 Uuh (293)		Ununoctium * 118 Uuo (294)

* The names and symbols for elements 113-116 and 118 are temporary. Final names will be selected when the elements' discoveries are verified.

Gadolinium 64 Gd 157.25	Terbium 65 Tb 158.93	Dysprosium 66 Dy 162.50	Holmium 67 Ho 164.93	Erbium 68 Er 167.26	Thulium 69 Tm 168.93	Ytterbium 70 Yb 173.05	Lutetium 71 Lu 174.97
Curium 96 Cm (247)	Berkelium 97 Bk (247)	Californium 98 Cf (251)	Einsteinium 99 Es (252)	Fermium 100 Fm (257)	Mendelevium 101 Md (258)	Nobelium 102 No (259)	Lawrencium 103 Lr (262)

SCIENCE SKILL HANDBOOK

MATH SKILL HANDBOOK

FOLDABLES HANDBOOK

REFERENCE HANDBOOK

GLOSSARY/ GLOSARIO

INDEX

Glossary/Glosario

Cómo usar el glosario en español:
1. Busca el término en inglés que desees encontrar.
2. El término en español, junto con la definición, se encuentran en la columna de la derecha.

Pronunciation Key

Use the following key to help you sound out words in the glossary.

a	back (BAK)	ew	food (FEWD)
ay	day (DAY)	yoo	pure (PYOOR)
ah	father (FAH thur)	yew	few (FYEW)
ow	flower (FLOW ur)	uh	comma (CAH muh)
ar	car (CAR)	u (+ con)	rub (RUB)
e	less (LES)	sh	shelf (SHELF)
ee	leaf (LEEF)	ch	nature (NAY chur)
ih	trip (TRIHP)	g	gift (GIHFT)
i (i + com + e)	idea (i DEE uh)	j	gem (JEM)
oh	go (GOH)	ing	sing (SING)
aw	soft (SAWFT)	zh	vision (VIH zhun)
or	orbit (OR buht)	k	cake (KAYK)
oy	coin (COYN)	s	seed, cent (SEED, SENT)
oo	foot (FOOT)	z	zone, raise (ZOHN, RAYZ)

English — A — Español

atom/Charles's Law — átomo/Ley de Charles

atom: a small particle that is the building block of matter. (p. 231)

átomo: partícula pequeña que es el componente básico de la materia. (pág. 231)

B

Boyle's Law: law that states that pressure of a gas increases if the volume decreases and pressure of a gas decreases if the volume increases, when temperature is constant. (p. 294)

Ley de Boyle: ley que afirma que la presión de un gas aumenta si el volumen disminuye y que la presión de un gas disminuye si el volumen aumenta, cuando la temperatura es constante. (pág. 294)

C

Charles's Law: law that states that volume of a gas increases with increasing temperature, if the pressure is constant. (p. 295)

Ley de Charles: ley que afirma que el volumen de un gas aumenta cuando la temperatura aumenta, si la presión es constante. (pág. 295)

chemical change: a change in matter in which the substances that make up the matter change into other substances with different chemical and physical properties. (p. 257)

chemical property: the ability or inability of a substance to combine with or change into one or more new substances. (p. 256)

compound: a substance containing atoms of two or more different elements chemically bonded together. (p. 234)

concentration: the amount of a particular solute in a given amount of solution. (p. 260)

condensation: the change of state from a gas to a liquid. (p. 286)

conduction: the transfer of thermal energy due to collisions between particles. (p. 206)

convection current: the movement of fluids in a cycle because of convection. (p. 211)

convection: the transfer of thermal energy by the movement of particles from one part of a material to another. (p. 210)

D

density: the mass per unit volume of a substance. (p. 243)

deposition: the process of changing directly from a gas to a solid. (p. 286)

dissolve: to form a solution by mixing evenly. (p. 235)

E

electric energy: energy carried by an electric current. (p. 165)

element: a substance that consists of only one type of atom. (p. 233)

energy: the ability to cause change. (p 161)

evaporation: the process of a liquid changing to a gas at the surface of the liquid. (p. 286)

cambio químico: cambio de la materia en el cual las sustancias que componen la materia se transforman en otras sustancias con propiedades químicas y físicas diferentes. (pág. 257)

propiedad química: capacidad o incapacidad de una sustancia para combinarse con o transformarse en una o más sustancias. (pág. 256)

compuesto: sustancia que contiene átomos de dos o más elementos diferentes unidos químicamente. (pág. 234)

concentración: cantidad de cierto soluto en una cantidad dada de solución. (pág. 260)

condensación: cambio de estado gaseoso a líquido. (pág. 286)

conducción: transferencia de energía térmica debido a colisiones entre partículas. (pág. 206)

corriente de convección: movimiento de fluidos en un ciclo debido a la convección. (pág. 211)

convección: transferencia de energía térmica por el movimiento de partículas de una parte de la materia a otra. (pág. 210)

densidad: cantidad de masa por unidad de volumen de una sustancia. (pág. 243)

deposición: proceso de cambiar directamente de gas a sólido. (pág. 286)

disolver: preparar una solución mezclando de manera homogénea. (pág. 235)

energía eléctrica: energía transportada por una corriente eléctrica. (pág. 165)

elemento: sustancia que consiste de un sólo tipo de átomo. (pág. 233)

energía: capacidad de causar cambio. (pág. 161)

evaporación: proceso por el cual un líquido cambia a gas en la superficie de dicho líquido. (pág. 286)

F

fossil fuel: the remains of ancient organisms that can be burned as an energy source. (p. 178)

friction: a contact force that resists the sliding motion of two surfaces that are touching. (p. 171)

combustible fósil: restos de organismos antiguos que pueden quemarse como fuente de energía. (pág. 178)

fricción: fuerza de contacto que resiste el movimiento de dos superficies que están en contacto. (pág. 171)

G

gas: matter that has no definite volume and no definite shape. (p. 278)

gas: materia que no tiene volumen ni forma definidos. (pág. 278)

H

heat engine: a machine that converts thermal energy into mechanical energy. (p. 218)

heat: the movement of thermal energy from a region of higher temperature to a region of lower temperature. (p. 201)

heating appliance: a device that converts electric energy into thermal energy. (p. 215)

heterogeneous mixture: a mixture in which substances are not evenly mixed. (p. 235)

homogeneous mixture: a mixture in which two or more substances are evenly mixed but not bonded together. (p. 235)

motor térmico: máquina que convierte energía térmica en energía mecánica. (pág. 218)

calor: movimiento de energía térmica de una región de alta temperatura a una región de baja temperatura. (pág. 201)

calentador: aparato que convierte energía eléctrica en energía térmica. (pág. 215)

mezcla heterogénea: mezcla en la cual las sustancias no están mezcladas de manera uniforme. (pág. 235)

mezcla homogénea: mezcla en la cual dos o más sustancias están mezcladas de manera uniforme, pero no están unidas químicamente. (pág. 235)

I

inexhaustible energy resource: an energy resource that cannot be used up. (p. 181)

recurso energético inagotable: recurso energético que no puede agotarse. (pág. 181)

K

kinetic (kuh NEH tik) energy: energy due to motion. (pp. 162, 282)

kinetic molecular theory: an explanation of how particles in matter behave. (p. 292)

energía cinética: energía debida al movimiento. (páges. 162, 282)

teoría cinética molecular: explicación de cómo se comportan las partículas en la materia. (pág. 292)

L

law of conservation of energy: law that states that energy can be transformed from one form to another, but it cannot be created or destroyed. (p. 170)

liquid: matter with a definite volume but no definite shape. (p. 276)

M

mass: the amount of matter in an object. (p. 242)

matter: anything that has mass and takes up space. (p. 231)

mechanical energy: sum of the potential energy and the kinetic energy in a system. (p. 165)

mixture: matter that can vary in composition. (p. 235)

N

nonrenewable energy resource: an energy resource that is available in limited amounts or that is used faster than it is replaced in nature. (p. 178)

nuclear energy: energy stored in and released from the nucleus of an atom. (p. 165)

P

physical change: a change in the size, shape, form, or state of matter that does not change the matter's identity. (p. 249)

physical property: a characteristic of matter that you can observe or measure without changing the identity of the matter. (p. 240)

potential (puh TEN chul) energy: stored energy due to the interactions between objects or particles. (p. 162)

pressure: the amount of force per unit area applied to an object's surface. (p. 293)

R

radiant energy: energy carried by an electro-magnetic wave. (p. 165)

ley de la conservación de la energía: ley que plantea que la energía puede transformarse de una forma a otra, pero no puede crearse ni destruirse. (pág. 170)

líquido: materia con volumen definido y forma indefinida. (pág. 276)

masa: cantidad de materia en un objeto. (pág. 242)

materia: cualquier cosa que tiene masa y ocupa espacio. (pág. 231)

energía mecánica: suma de la energía potencial y la energía cinética en un sistema. (pág. 165)

mezcla: materia cuya composición puede variar. (pág. 235)

recurso energético no renovable: recurso energético disponible en cantidades limitadas o que se usa más rápido de lo que se repone en la naturaleza. (pág. 178)

energía nuclear: energía almacenada en y liberada por el núcleo de un átomo. (pág. 165)

cambio físico: cambio en el tamaño, la forma o el estado de la materia en el que no cambia la identidad de la materia. (pág. 249)

propiedad física: característica de la materia que puede observarse o medirse sin cambiar la identidad de la materia. (pág. 240)

energía potencial: energía almacenada debido a las interacciones entre objetos o partículas. (pág. 162)

presión: cantidad de fuerza por unidad de área aplicada a la superficie de un objeto. (pág. 293)

energía radiante: energía que transporta una onda electromagnética. (pág. 165)

SCIENCE SKILL HANDBOOK

MATH SKILL HANDBOOK

REFERENCE HANDBOOK

GLOSSARY/ GLOSARIO

INDEX

Glossary/Glosario • **G-5**

radiation: the transfer of thermal energy by electromagnetic waves. (p. 205)

refrigerator: a device that uses electric energy to pump thermal energy from a cooler location to a warmer location. (p. 216)

renewable energy resource: an energy resource that is replaced as fast as, or faster than, it is used. (p. 180)

radiación: transferencia de energía térmica por ondas electromagnéticas. (pág. 205)

refrigerador: aparato que usa energía eléctrica para bombear energía térmica desde un lugar más frío hacia uno más caliente. (pág. 216)

recurso energético renovable: recurso energético que se repone tan rápido, o más rápido, de lo que se consume. (pág. 180)

S

solid: matter that has a definite shape and a definite volume. (p. 275)

solubility: the maximum amount of solute that can dissolve in a given amount of solvent at a given temperature and pressure. (p. 244)

sound energy: energy carried by sound waves. (p. 165)

specific heat: the amount of thermal energy it takes to increase the temperature of 1 kg of a material by 1ºC. (p. 207)

sublimation: the process of changing directly from a solid to a gas. (p. 286)

substance: matter with a composition that is always the same. (p. 233)

surface tension: the uneven forces acting on the particles on the surface of a liquid. (p. 277)

sólido: materia con forma y volumen definidos. (pág. 275)

solubilidad: cantidad máxima de soluto que puede disolverse en una cantidad dada de solvente a temperatura y presión dadas. (pág. 244)

energía sonora: energía que transportan las ondas sonoras. (pág. 165)

calor específico: cantidad de energía térmica necesaria para aumentar la temperatura de 1 Kg de un material en 1°C. (pág. 207)

sublimación: proceso de cambiar directamente de sólido a gas. (pág. 286)

sustancia: materia cuya composición es siempre la misma. (pág. 233)

tensión superficial: fuerzas desiguales que actúan sobre las partículas en la superficie de un líquido. (pág. 277)

T

temperature: the measure of the average kinetic energy of the particles in a material. (pp. 199, 282)

thermal conductor: a material through which thermal energy flows quickly. (p. 206)

thermal contraction: a decrease in a material's volume when the temperature is decreased. (p. 208)

thermal energy: the sum of the kinetic energy and the potential energy of the particles that make up an object. (pp. 165, 198, 283)

thermal expansion: an increase in a material's volume when the temperature is increased. (p. 208)

temperatura: medida de la energía cinética promedio de las partículas de un material. (páges. 199, 282)

conductor térmico: material mediante el cual la energía térmica se mueve con rapidez. (pág. 206)

contracción térmica: disminución del volumen de un material cuando disminuye la temperatura. (pág. 208)

energía térmica: suma de la energía cinética y potencial de las partículas que componen un objeto. (páges. 165, 198, 283)

expansión térmica: aumento en el volumen de un material cuando aumenta la temperatura. (pág. 208)

SCIENCE SKILL HANDBOOK

MATH SKILL HANDBOOK

REFERENCE HANDBOOK

GLOSSARY/ GLOSARIO

INDEX

thermal insulator: a material through which thermal energy flows slowly. (p. 206)

thermostat: a device that regulates the temperature of a system. (p. 216)

aislante térmico: material en el cual la energía térmica se mueve con lentitud. (pág. 206)

termostato: aparato que regula la temperatura de un sistema. (pág. 216)

vapor: the gas state of a substance that is normally a solid or a liquid at room temperature. (p. 278)

vaporization: the change in state from a liquid to a gas. (p. 285)

viscosity (vihs KAW sih tee): a measurement of a liquid's resistance to flow. (p. 276)

vapor: estado gaseoso de una sustancia que normalmente es sólida o líquida a temperatura ambiente. (pág. 278)

vaporización: cambio de estado líquido a gaseoso. (pág. 285)

viscosidad: medida de la resistencia de un líquido a fluir. (pág. 276)

work: the amount of energy used as a force moves an object over a distance. (p. 162)

trabajo: cantidad de energía usada como fuerza que mueve un objeto a cierta distancia. (pág. 162)

Index

A

Academic Vocabulary, 207, 234, 292
Air. *See also* **Atmosphere**
  as thermal insulator, 213
Atmosphere. *See also* **Air**
  convection currents in, 211, *211*
Atom(s)
  combinations and arrangements
    of, 233
  in compounds, 234, *234*
  in elements, 233, *233, 234*
  explanation of, **231,** 241
  parts of, 232, *232,* 232 *lab*
  in thermal conductors, 206
Average kinetic energy
  temperature as, 199, 206, 208

B

Big Idea, 158, 188, 194, 222, 228, 264,
  270, 300
  Review, 191, 225, 267, 303
Bimetallic coil(s)
  explanation of, *216,* **216**
Biomass
  advantages and disadvantages of,
    184
  explanation of, 182
Boiling, 285, *285*
Boiling point
  explanation of, 243, *243, 245,* 250
  of water, 287
Bond(s), 245
Boyle, Robert, 294
Boyle's law, 274, 294, *294*
Bulb thermometer(s), 200

C

Carbon
  particle arrangements in, 275, *275*
Carbon dioxide
  in atmosphere, 167, 179
  explanation of, 234, *234*
Cavity wall(s), 213
Cell phones
  radiant energy from, 172, *172*
Celsius scale, 200, *200*
Chapter Review, 190–191, 224–225,
  266–267, 302–303
Charles, Jocque, 295
Charles's law, 295–296, *296*
Chemical change(s)
  clues of, 258 *lab*
  explanation of, *257,* **257,** 258
Chemical energy
  use of, 172, *173*

Chemical equation(s)
  balancing, 259, *259*
  explanation of, 258, *258*
Chemical formula(s), 234, 236, 258,
  258, 259
Chemical potential energy
  explanation of, 163, *163,* 283
  in fossil fuels, 179
Chemical property(ies), 256
Chemical reaction(s)
  explanation of, 258
  rate of, 260, *260*
Coal
  formation of, 179
Coin(s), 238
Common Use. *See* **Science Use v.**
  Common Use
Compound(s)
  explanation of, **234,** *236*
  properties of, 234
  solutions v., 236
Compressor(s), 217
Concentration
  rate of chemical reactions and, 260,
    260
Concrete
  expansion in, 208, *208*
Condensation, 251, **286**
Conduction, 206
Conductivity, 244, *244*
Conservation
  of mass and energy, 288, *288*
Container(s)
  insulated, 220–221 *lab*
Control joint(s), 208, *208*
Convection, 210
Convection current(s), 211, *195*
Coolant(s), 217
Critical thinking, 166, 174, 185, 191
Cubic centimeter(s) (cm³), 273
Cubic meter(s) (m³), 273

D

Dam(s), 181, *181*
Density
  explanation of, **243,** *245,* 273
  thermal contraction and thermal
    expansion and, 210
Deposition, 251, *251,* **286**
Dissolve, 235
Dissolving, 252, *252*
Dry ice, 286, *286*

E

Elastic potential energy, 163, *163*
Electric compressors, 217

Electric energy
  converted to thermal energy, 215
Electric power plant(s)
  function of, 178, *178*
Electrical conductivity, 244
Electrical energy
  changed to radiant energy, 169, *169*
  explanation of, *165,* **165**
  kinetic energy transformed into, 181
  nuclear energy transformed to, 180,
    180
  sources of, 178, *179,* 183 *lab*
  use of, 173
Electron(s)
  explanation of, 232, *232*
  transfer of kinetic energy by, 206
Electronic thermometer(s), 200
Element(s)
  atoms in, 233, *233, 234*
  explanation of, **233,** 236
  periodic table of, 233, *233*
Energy
  changes between forms of, 169–170,
    169 *lab,* 177
  conservation of, 288
  from electric power plants, 178, *178*
  electric, 215
  explanation of, **161**
  forms of, 164, 165, *165,* 173 *lab*
  kinetic, 162, 197–199, 206, 208, 282
  law of conservation of, **170**–171
  mechanical, 197, 218
  potential, 162–163, 197, 198, *198,*
    283, *283*
  sources of, 177, *177,* 183
  thermal, 198, 199, 201, 203, *205,*
    205–211, *206, 207, 208, 209, 210,*
    211, 215–218, *216, 217, 218,* 283,
    284, 285, 287, 287
  transformations of, 175, *175*
  use of, 172–173
  work and, 164
Energy resource(s)
  advantages and disadvantages of,
    184
  conservation of, 183
  differences between, 177 *lab*
  explanation of, **159**
  inexhaustible, 181
  nonrenewable, 178–180, *179, 180,*
    183
  renewable, 180–182, *181, 182*
Engine(s)
  heat, 218, **218**
Evaporation, 286
Experiment(s)
  design of, 262–263 *lab,* 298–299 *lab*
Extrapolated, 296

F

Fahrenheit scale, 200, *200*
Foldables, 162, 172, 178, 189, 199, 206, 215, 223, 236, 241, 249, 257, 265, 274, 293, 301
Food(s)
    freeze-dried, 280
Fossil fuel(s)
    advantages and disadvantages of, *184*
    CO_2 levels and, 167
    explanation of, **178**
    formation of, 179, *179*
    global warming and, 167, 179
    supply of, 183
    use of, 179
Freeze-dried food(s), 280
Freezing, 251, *251,* 284
Friction, **159,** **171**

G

Gas(es)
    behavior of, 292
    Boyle's law and, 294, *294*
    changes of liquids to/from, *285,* 285–286
    changes of solids to/from, 286, *286*
    Charles's law and, 295–296, *296*
    condensation and, 251
    explanation of, 241, *241,* **278**
    forces between, 278
    particles in, 278
    pressure and volume of, 292, 294–295
    temperature and volume of, 295, *295*
Gasoline
    energy from, 218
Geothermal energy
    advantages and disadvantages of, *184*
    explanation of, 182
Glass
    ovenproof, 209
Global warming
    fossil fuels and, 167, 179
Gram(s) (g), 273
Gravitational potential energy
    explanation of, 163, *163,* 197
    transformed into kinetic energy, 175, *175*
Gravity, 197
Green Science, 167
Greenhouse gas(es)
    carbon dioxide as, 167

H

Heat
    explanation of, **201**
    specific, 207, *207*
    thermal energy and, 201, 209 *lab*
Heat engine(s), *218,* **218**
Heating appliance(s), **215**
Heterogeneous mixture(s), 235, *235, 236*

Home(s)
    thermal insulators in, 213
Homogeneous mixture(s)
    explanation of, 235, *235, 236*
    salt water as, 236
Hot-air balloon(s), 209, *209*
How It Works, 238, 280
Hydroelectric power plant(s)
    advantages and disadvantages of, *184*
    explanation of, 181, *181*

I

Ice
    dry, 286, *286*
    melting of, 287
Inexhaustible energy resource(s), **181,** 182
Insulated container(s), 220–221 *lab*
Interpret Graphics, 166, 174, 185

K

Kelvin scale, 200, *200*
Key Concepts, 160, 168, 176
    Check, 161, 162, 163, 164, 165, 170, 171, 173, 178, 180, 183
    Summary, 188
    Understand, 166, 174, 190
Kilogram(s) (kg), 273
Kinetic energy
    changes between potential energy and, 170, *170*
    explanation of, 162, *162,* **197,** 198, **282**
    gravitational potential energy transformed into, 175, *175*
    temperature as average, 199, 206, 208
    transformed into electrical energy, 181
Kinetic molecular theory, **292**

L

Lab, 186–187, 220–221, 262–263, 298–299. *See also* **Launch Lab; MiniLab; Skill Practice**
Launch Lab, 161, 169, 177, 197, 215, 231, 240, 249, 256, 273, 282, 292
Law of conservation of energy
    explanation of, **170,** 177
    friction and, 171, *171*
Light energy. *See* **Radiant energy**
Lesson Review, 166, 174, 185, 202, 212, 219, 237, 246, 253, 261, 279, 289, 297
Liquid(s)
    changes of gases to/from, *285,* 285–286
    changes of solids to/from, 284, *284*
    explanation of, 241, *241,* **276**
    particles in, 276, 282 *lab*
    surface tension in, 277, *277*
    viscosity of, 276
Liter(s) (L), 273
Lyophilizaton, 280

M

Magnetism, *245*
Mass
    conservation of, 252, *252,* 259, 288
    explanation of, 242, *242,* 242 *lab,* *244,* 273
    kinetic energy and, 162, *162*
Math Skills, 180, 185, 191, 200, 202, 225, 243, 246, 267, 294, 297, 303
Matter. *See also* **Gases; Liquids; Plasma; Solids;** States of matter
    classification of, 232, 233, 236, *236*
    explanation of, **231, 273**
    mass of, 242, *242*
    methods to describe, 273
    observations of, 161
    particles in, 274, *274*
    physical changes in, *249,* 249–252, *250, 251, 252*
    physical properties of, 240–245, *241, 242, 243, 244*
    states of, 241, *241, 245, 250,* 250–252, *251, 252*
    volume of, 243, *243*
Mechanical energy
    explanation of, *165,* **165,** 197
    thermal energy converted to, 218
Melting, 284
Melting point
    explanation of, 243
    for solids, 284
    of water, 287
Metal(s)
    in coins, 238
Milliliter(s) (mL), 243, 273
MiniLab, 164, 173, 183, 201, 209, 217, 232, 242, 251, 277, 288, 295. *See also* **Lab**
Mixture(s)
    explanation of, **235,** *236*
    heterogeneous, 235, *235, 236*
    homogeneous, 235, *235,* 236, *236*
    separation of, 245, *245*
Molecule(s)
    surface tension in liquid, 277
Motion
    particles in, 274
    random, 274
Motor vehicle(s)
    engines in, 218, *218*

N

Natural gas
    formation of, 179
    use of, 179
Neutron(s), 232, *232*
Nonrenewable energy resource(s)
    advantages and disadvantages of, *184*
    explanation of, **178**
    fossil fuels as, 178–179, *179*
    nuclear energy as, 180, *180*
    percentage of energy from, *183*

SCIENCE SKILL HANDBOOK

MATH SKILL HANDBOOK

REFERENCE HANDBOOK

GLOSSARY/ GLOSARIO

INDEX

Nuclear energy
advantages and disadvantages of, *184*
explanation of, **165,** *165,* 180
transformed to electrical energy, 180, *180*
Nuclear power plant(s), 180, *180*
Nuclear waste
storage of, 180
Nuclei, 180
Nucleus, 232, *232,* 233

O

Ovenproof glass, 209

P

Particle(s)
forces between, 274, *274*
in gas, 278
in liquids, 276, 282 *lab*
in motion, 274
in solids, 275, *275*
Petroleum
formation of, 179
use of, 179
Physical change(s)
explanation of, **249**
in shape and size, 249
in state of matter, *250,* 250–252, *251, 252,* 256 *lab*
Physical property(ies)
explanation of, **240**
size-dependent, *242,* 242–243
size-independent, *243,* 243–245, *244, 245*
states of matter and, 241, *241*
Plasma, 273
Potential energy
changes between kinetic energy and, 170, *170,* 175, *175*
explanation of, **159, 162, 197,** 198, *198,* 283, *283*
types of, 163, *163*
work and, *164*
Pressure
explanation of, **293**
of gas, 292 *lab,* 294–295
relationship between volume and, 293
Property(ies)
chemical, 256, 258 *lab*
comparison of, 256
explanation of, **240**
physical, 240–245, *241, 242, 243, 244, 245*
Proton(s), 232, *232,* 233

R

Radiant energy
explanation of, *165,* **165,** *169,* **169**
use of, 172, *172*
Radiation, 205
Random motion, 274

Reaction(s), 258
Reading Check, 163, 170, 173, 179, 180
Refrigerator(s)
coolants in, 217, *217*
explanation of, **216**
Renewable energy resource(s)
advantages and disadvantages of, *184*
biomass as, 182
explanation of, **180**
geothermal energy as, 182, *182*
hydroelectric power plants and, 181, *181*
solar energy as, 181, *181*
wind energy as, 182, *182*
Reservoir, 181, *181*
Review Vocabulary, 180, 197, 240, 273. *See also* **Vocabulary**
Rock(s), 231, *231*

S

Salt water
composition of, 236
Sand
mixture of water and, 235
Science & Society, 213
Science Methods, 187, 221, 263, 299
Science Use v. Common Use, 169, 205, 245, 286. *See also* **Vocabulary**
Shelter(s)
thermal insulators in, 213
Skill Practice, 175, 203, 247, 254, 290. *See also* **Lab**
Solar energy
advantages and disadvantages of, *184*
explanation of, 181, *181*
Solid(s)
changes of gases to/from, 286, *286*
changes of liquid to/from, 284, *284*
explanation of, 241, *241,* **275**
particles in, 275, *275*
thermal energy added to, 250
types of, 275, *275*
Solubility, 244, *245*
Solute(s), 235, 236
Solution(s)
compounds v., 236
explanation of, 235
Solvent(s), 235, 236
Sound energy, *165,* **165**
Specific, 207
Specific heat, 207
Speed
kinetic energy and, 162, *162*
Standardized Test Practice, 192–193, 226–227, 268–269, 304–305
States of matter. *See also* **Gases; Matter; Plasma; Solids**
adding thermal energy to, 250, *250*
dissolving and, 252, *252*
explanation of, 241, *241*
removal of, 251, *251*
Study Guide, 188–189, 222–223, 264–265, 300–301
Sublimation, 250, **286**

Substance(s)
compounds as, 234, *234,* 236
elements as, 233, *233*
explanation of, **233**
identification of, 240 *lab,* 254
mixtures as combinations of, 235, *235,* 245
Sun
energy from, 177, *177*
thermal energy from, 205, *205,* 206
Surface area
rate of chemical reactions and, 260, *260*
Surface tension, 277, 277 *lab*

T

Temperature
accuracy of touch to predict, 205 *lab*
average kinetic energy and, 199, 206, 208
change in state of matter due to, 250, *250*
explanation of, **199, 282**
measurement of, 197 *lab,* 200, *200,* 201
rate of chemical reactions and, 260, *260*
thermal energy and, 199
thermal energy v., *284, 285,* 296
volume and, 296 *lab*
Temperature scale(s), 200, *200,* 7*lab*
Thermal conductor(s)
explanation of, **206**
specific heat in, 207, *207*
Thermal contraction
explanation of, **208,** 210
in hot-air balloons, 209, *209*
Thermal energy
addition of, 287
change in state of matter due to, 250, *250,* 251, *251*
chemical potential energy changed to, 179
explanation of, *165,* **165, 198,** 199, 215, **283**
friction and, 171, *171*
heat and, 201, 209 *lab*
heat engines and, 218, *218*
heating appliances and, 215
inside Earth, 182
refrigerators and, 216–217, *217*
removal of, 287, *287,* 290
temperature and, 199
temperature v., *284, 285,* 296
thermostats and, 216, *216*
use of, 172
work and, 217 *lab*
Thermal energy transfer
conduction and, *206,* 206–207, *207*
convection and, *210,* 210–211, *211*
effect of materials on, 203
explanation of, 205
radiation and, 205, *205*
thermal expansion and contraction and, *208,* 208–209, *209*

Thermal expansion
explanation of, **208,** 210
in hot-air balloons, 209, *209*
Thermal insulator(s)
explanation of, **206**
in shelters, 213
Thermometer
explanation of, 200, *200*
water, 288 *lab*
Thermostat(s), *216,* **216**

U.S. Mint, 238
Unique, 234

Vacuum, 205
Vapor, 278, 286
Viscosity, 276
Visual Check, 170, 178, 181

Vocabulary, 159, 160, 168, 176. *See also*
**Review Vocabulary; Science
Use v. Common Use; Word
Origin**
Use, 166, 174, 185, 189
Volume
explanation of, 243, *244,* 273
gas, 292 *lab,* 293–295
liquid, 273
temperature and, 295 *lab*

Waste energy
use of, 173, *173*
Water
boiling point of, 287
composition of, 236
freezing point of, 290
heating curve of, 287, *287*
melting point of, 287
specific heat of, 207
states of, 287, *287*

Weight
mass and, 242, 242 *lab*
What do you think?, 166, 174, 185
Wind energy
advantages and disadvantages of,
184
explanation of, 182, *182*
Wind turbine(s)
explanation of, 182, *182*
observation of, 186–187 *lab*
Word Origin, 161, 171, 178, 199, 210,
216, 231, 243, 249, 257, 273, 286.
See also **Vocabulary**
Work
energy and, 164, *164*
thermal energy and, 217 *lab*
Writing In Science, 191, 225, 267, 303

SCIENCE SKILL HANDBOOK

MATH SKILL HANDBOOK

REFERENCE HANDBOOK

GLOSSARY/ GLOSARIO

INDEX

Credits

Photo Credits

PERIODIC TABLE OF THE ELEMENTS

	Element	Hydrogen
	Atomic number	1
	Symbol	H
	Atomic mass	1.01

State of matter

🎈 Gas
💧 Liquid
⬜ Solid
⊙ Synthetic

1

1
Hydrogen
1
H 🎈
1.01

2

A column in the periodic table is called a **group**.

	1	2
1	Hydrogen 1 **H** 🎈 1.01	
2	Lithium 3 **Li** ⬜ 6.94	Beryllium 4 **Be** ⬜ 9.01
3	Sodium 11 **Na** ⬜ 22.99	Magnesium 12 **Mg** ⬜ 24.31

	3	4	5	6	7	8	9
4	Potassium 19 **K** ⬜ 39.10	Calcium 20 **Ca** ⬜ 40.08	Scandium 21 **Sc** ⬜ 44.96	Titanium 22 **Ti** ⬜ 47.87	Vanadium 23 **V** ⬜ 50.94	Chromium 24 **Cr** ⬜ 52.00	Manganese 25 **Mn** ⬜ 54.94
5	Rubidium 37 **Rb** ⬜ 85.47	Strontium 38 **Sr** ⬜ 87.62	Yttrium 39 **Y** ⬜ 88.91	Zirconium 40 **Zr** ⬜ 91.22	Niobium 41 **Nb** ⬜ 92.91	Molybdenum 42 **Mo** ⬜ 95.96	Technetium 43 **Tc** ⊙ (98)
6	Cesium 55 **Cs** ⬜ 132.91	Barium 56 **Ba** ⬜ 137.33	Lanthanum 57 **La** ⬜ 138.91	Hafnium 72 **Hf** ⬜ 178.49	Tantalum 73 **Ta** ⬜ 180.95	Tungsten 74 **W** ⬜ 183.84	Rhenium 75 **Re** ⬜ 186.21
7	Francium 87 **Fr** ⬜ (223)	Radium 88 **Ra** ⬜ (226)	Actinium 89 **Ac** ⬜ (227)	Rutherfordium 104 **Rf** ⊙ (267)	Dubnium 105 **Db** ⊙ (268)	Seaborgium 106 **Sg** ⊙ (271)	Bohrium 107 **Bh** ⊙ (272)

Iron 26 **Fe** ⬜ 55.85 — Cobalt 27 **Co** ⬜ 58.93

Ruthenium 44 **Ru** ⬜ 101.07 — Rhodium 45 **Rh** ⬜ 102.91

Osmium 76 **Os** ⬜ 190.23 — Iridium 77 **Ir** ⬜ 192.22

Hassium 108 **Hs** ⊙ (270) — Meitnerium 109 **Mt** ⊙ (276)

The number in parentheses is the mass number of the longest lived isotope for that element.

A row in the periodic table is called a **period**.

	Cerium	Praseodymium	Neodymium	Promethium	Samarium	Europium
Lanthanide series	58 **Ce** ⬜ 140.12	59 **Pr** ⬜ 140.91	60 **Nd** ⬜ 144.24	61 **Pm** ⊙ (145)	62 **Sm** ⬜ 150.36	63 **Eu** ⬜ 151.96

	Thorium	Protactinium	Uranium	Neptunium	Plutonium	Americium
Actinide series	90 **Th** ⬜ 232.04	91 **Pa** ⬜ 231.04	92 **U** ⬜ 238.03	93 **Np** ⊙ (237)	94 **Pu** ⊙ (244)	95 **Am** ⊙ (243)